# DISCIPLINE-BASED EDUCATION RESEARCH

## Understanding and Improving Learning in Undergraduate Science and Engineering

Committee on the Status, Contributions, and Future Directions of
Discipline-Based Education Research

Board on Science Education

Division of Behavioral and Social Sciences and Education

Susan R. Singer, Natalie R. Nielsen,
and Heidi A. Schweingruber, *Editors*

NATIONAL RESEARCH COUNCIL
*OF THE NATIONAL ACADEMIES*

THE NATIONAL ACADEMIES PRESS
Washington, D.C.
**www.nap.edu**

THE NATIONAL ACADEMIES PRESS    500 Fifth Street, NW    Washington, DC 20001

NOTICE: The project that is the subject of this report was approved by the Governing Board of the National Research Council, whose members are drawn from the councils of the National Academy of Sciences, the National Academy of Engineering, and the Institute of Medicine. The members of the committee responsible for the report were chosen for their special competences and with regard for appropriate balance.

This study was supported by Contract/Grant No. 0934453 between the National Academy of Sciences and the National Science Foundation. Any opinions, findings, conclusions, or recommendations expressed in this publication are those of the author(s) and do not necessarily reflect the views of the organizations or agencies that provided support for the project.

### Library of Congress Cataloging-in-Publication Data

National Research Council (U.S.). Committee on the Status, Contributions, and Future Directions of Discipline-Based Education Research.
  Discipline-based education research : understanding and improving learning in undergraduate science and engineering / Susan R. Singer, Natalie R. Nielsen, and Heidi A. Schweingruber, editors. ; Committee on the Status, Contributions, and Future Directions of Discipline-Based Education Research, Board on Science Education, Division of Behavioral and Social Sciences and Education, National Research Council of the National Academies.
      pages cm
  Includes bibliographical references.
  ISBN 978-0-309-25411-3 (paperback) — ISBN 0-309-25411-6 (paperback)  1. Science—Study and teaching (Secondary)—United States. 2. Engineering—Study and teaching (Secondary)—United States. 3. Science—Study and teaching (Higher)—United States. 4. Engineering—Study and teaching (Higher)—United States. 5. Universities and colleges—Curricula—United States.  I. Singer, Susan R., editor. II. Nielsen, Natalie R., editor. III. Schweingruber, Heidi A., editor. IV. Title.
  Q183.3.A1N3585 2012
  507.1'173—dc23
                              2012027266

Additional copies of this report are available from the National Academies Press, 500 Fifth Street, NW, Keck 360, Washington, DC 20001; (800) 624-6242 or (202) 334-3313; http://www.nap.edu.

Suggested citation: National Research Council. (2012). *Discipline-Based Education Research: Understanding and Improving Learning in Undergraduate Science and Engineering.* S.R. Singer, N.R. Nielsen, and H.A. Schweingruber, Editors. Committee on the Status, Contributions, and Future Directions of Discipline-Based Education Research. Board on Science Education, Division of Behavioral and Social Sciences and Education. Washington, DC: The National Academies Press.

# THE NATIONAL ACADEMIES
*Advisers to the Nation on Science, Engineering, and Medicine*

The **National Academy of Sciences** is a private, nonprofit, self-perpetuating society of distinguished scholars engaged in scientific and engineering research, dedicated to the furtherance of science and technology and to their use for the general welfare. Upon the authority of the charter granted to it by the Congress in 1863, the Academy has a mandate that requires it to advise the federal government on scientific and technical matters. Dr. Ralph J. Cicerone is president of the National Academy of Sciences.

The **National Academy of Engineering** was established in 1964, under the charter of the National Academy of Sciences, as a parallel organization of outstanding engineers. It is autonomous in its administration and in the selection of its members, sharing with the National Academy of Sciences the responsibility for advising the federal government. The National Academy of Engineering also sponsors engineering programs aimed at meeting national needs, encourages education and research, and recognizes the superior achievements of engineers. Dr. Charles M. Vest is president of the National Academy of Engineering.

The **Institute of Medicine** was established in 1970 by the National Academy of Sciences to secure the services of eminent members of appropriate professions in the examination of policy matters pertaining to the health of the public. The Institute acts under the responsibility given to the National Academy of Sciences by its congressional charter to be an adviser to the federal government and, upon its own initiative, to identify issues of medical care, research, and education. Dr. Harvey V. Fineberg is president of the Institute of Medicine.

The **National Research Council** was organized by the National Academy of Sciences in 1916 to associate the broad community of science and technology with the Academy's purposes of furthering knowledge and advising the federal government. Functioning in accordance with general policies determined by the Academy, the Council has become the principal operating agency of both the National Academy of Sciences and the National Academy of Engineering in providing services to the government, the public, and the scientific and engineering communities. The Council is administered jointly by both Academies and the Institute of Medicine. Dr. Ralph J. Cicerone and Dr. Charles M. Vest are chair and vice chair, respectively, of the National Research Council.

**www.national-academies.org**

*v*

# Acknowledgments

This report is made possible by the important contributions of National Research Council (NRC) leadership and staff, and many other organizations. First, we acknowledge the sponsorship of the National Science Foundation. We particularly thank Myles Boylan, a program director in the Division of Undergraduate Education, who supported and encouraged the development of the report.

This study had its origins in two workshops in 2008 on promising practices in undergraduate science, technology, engineering, and mathematics. The committee thanks Carl Wieman, who was the chair of the Board on Science Education (BOSE) at the time, for his vision and leadership to help turn those workshops into this consensus study. We also thank Heidi Schweingruber and Margaret Hilton for serving as the institutional memory and helping to build as many connections as possible between those workshops and this study.

Over the course of this study, members of the committee benefited from discussion and presentations by the many individuals who participated in our four fact-finding meetings. We acknowledge the efforts of the 22 authors who prepared background papers. Janelle Bailey (University of Nevada, Las Vegas), George Bodner (Purdue University), Karen Cummings (Southern Connecticut State University), Robert DeHaan (Emory University), and Jack Lohmann (Georgia Institute of Technology) with Jeffrey Froyd (Texas A&M University) were asked to describe the developmental histories of education research in astronomy, chemistry, physics, biology, and engineering, respectively. We also commissioned literature reviews of research on teaching and learning from Janelle Bailey (University of Nevada, Las Vegas) in astronomy, Clarissa Dirks (The Evergreen State College) in biology, Jennifer Docktor and Jose Mestre (both of University of Illinois at Urbana–Champaign) in physics, Michael Piburn (Arizona State University), Kaatje van der Hoeven Kraft (Mesa Community College) and Heather Pacheco (Arizona State University) in the geosciences, Marilla Svinicki (University of Texas) in engineering, and Marcy Towns (Purdue University) with Adam Kraft in chemistry. To facilitate our examination across disciplines and into cognitive science, Elliott Sober (University of Wisconsin–Madison) was asked to prepare a paper on epistemological similarities and differences in the sciences. Richard Mayer (University of California, Santa Barbara) prepared a paper that applied the science of learning to undergraduate science education, and Mary Hegarty (University of California, Santa Barbara) prepared a paper on spatial thinking and the use of representations in the sciences. Ann Austin (Michigan State University) prepared a paper on the factors that influence faculty members' instructional decision making, and Noah Finkelstein (University of Colorado) with Julie Libarkin (Michigan State University) analyzed the role of the National Science Foundation's Postdoctoral Fellowship in Science, Mathematics, Engineering, and Technology Program as a pathway for discipline-based education researchers. At our fourth meeting, Kathy Perkins (University of Colorado) discussed the University of Colorado's Science Education Initiative and its role in advancing discipline-based education research.

We also are deeply grateful to the many individuals at the NRC who assisted the committee. The success of a consensus study such as this report involves the efforts of countless NRC staff members who work behind the scenes. In particular, this report would not have been possible without Heidi Schweingruber and Margaret Hilton, who helped to shape the meeting agendas and the project's overall trajectory, and participated in committee deliberations. They also made profound contributions to the report by editing individual chapters, providing feedback on the report as a whole, participating in regular meetings with the committee chair, and generally making themselves available to provide advice and guidance. We are grateful to Anthony Brown, Dorothy Majewski, and Rebecca Krone, who arranged logistics for our meetings and facilitated the proceedings of the meetings themselves. We also appreciate Anthony Brown's assiduous efforts to compile the reference lists in each chapter of the report. And we thank Kirsten Sampson Snyder, who shepherded the report through the NRC review process; Amy Smith, who edited the draft report; and Yvonne Wise for processing the report through final production.

The report also benefitted from the contributions of several affiliates of the NRC. Rochelle Urban and Kristina Mitchell, participants in the NRC's Christine Mirzayan Science and Technology Policy Graduate Fellowship Program, provided research support. Tom Foster, a physics education researcher who spent his sabbatical from Southern Illinois

University–Edwardsville as a consultant to the NRC, provided valuable assistance during the report writing.

This report has been reviewed in draft form by individuals chosen by their diverse perspective and technical expertise, in accordance with procedures approved by the NRC's Report Review Committee. The purpose of this independent review is to provide candid and critical comments that will assist the institution in making its published report as sound as possible and to ensure that the report meets institutional standards for objectivity, evidence, and responsiveness to the study charge. The review comments and draft manuscript remain confidential to protect the integrity of the deliberative process. We thank the following individuals for their review of this report: Alice M. Agogino, Mechanical Engineering, University of California, Berkeley; Mark R. Connolly, Wisconsin Center for Education Research, University of Wisconsin–Madison; Roger M. Downs, Department of Geography, Pennsylvania State University; Mark Dynarski, Pemberton Research Associates, LLC, East Windsor, NJ; Gary Gladding, Department of Physics, University of Illinois at Urbana–Champaign; Paula Heron, Department of Physics, University of Washington; Thomas Holme, Chemistry Department, Iowa State University; C. Judson King, Department of Chemical and Biomolecular Engineering (Emeritus), University of California, Berkeley; Michael Klymkowsky, Department of Molecular, Cellular and Developmental Biology, University of Colorado at Boulder; Joseph Krajcik, College of Natural Science and College of Education, University of Michigan; Cathryn A. Manduca, Science Education Resource Center, Carleton College; Robert D. Mathieu, Department of Astronomy, University of Wisconsin–Madison; Lindsey Richland, Department of Comparative Human Development, University of Chicago; Lorrie A. Shepard, School of Education, University of Colorado at Boulder; and Vicente Talanquer, Department of Chemistry and Biochemistry, University of Arizona.

Although the reviewers listed above have provided constructive comments and suggestions, they are not asked to endorse the conclusions or recommendations nor did they see the final draft of the report before its release. Susan Hanson of Clark University and Adam Gamoran of the University of Wisconsin–Madison oversaw the review of this report. Appointed by the NRC, they were responsible for making certain that an independent examination of this report was carried out in accordance with institutional procedures and that all review comments were carefully considered. Responsibility for the final content of this report rests entirely with the authoring committee and the institution.

Susan R. Singer, *Chair*
Committee on the Status, Contributions, and Future Directions of
Discipline-Based Education Research
Natalie R. Nielsen, *NRC Study Director*

# Contents

## SECTION II. CONTRIBUTIONS OF DISCIPLINE-BASED EDUCATION RESEARCH

In Memoriam
Michael Edward Martinez, 1956-2012


*A scholar in cognition, intelligence, and science and mathematics
learning, Michael Martinez brought an expansive mind, good
humor, humility, and a consensus-building approach to our
committee and many other National Research Council activities.
His diverse career path included teaching high school science;
developing computer-based assessments in science, architecture,
and engineering at the Educational Testing Service; serving as a
Fulbright Scholar in the Fiji Islands; and managing the National
Science Foundation's role in the Interagency Educational
Research Initiative. He made great contributions in all of these
capacities, and this study was no exception. We benefited
enormously from his work with us, and we were deeply saddened
by his death shortly before this report was released. We dedicate
this book to a wonderful scholar, intellect, and friend.*

# Executive Summary

The United States faces a great imperative to improve undergraduate science and engineering education. Preparing a diverse technical workforce and science-literate citizenry will require significant changes to undergraduate science and engineering education. These changes include supporting an emerging, interdisciplinary research enterprise that combines the expertise of scientists and engineers with methods and theories that explain learning. This enterprise, discipline-based education research (DBER), investigates learning and teaching in a discipline from a perspective that reflects the discipline's priorities, worldview, knowledge, and practices. Informed by and complementary to research on learning and cognition, DBER already has generated insights that can be used to better prepare students to understand and address current and future societal challenges.

Recognizing DBER's emergence as a vital area of scholarship and its potential to improve undergraduate science and engineering education, the National Science Foundation requested that the National Research Council convene the Committee on the Status, Contributions, and Future Directions of Discipline-Based Education Research to conduct a synthesis study of DBER. Looking across physics, chemistry, engineering, biology, the geosciences, and astronomy, the committee's charge was to

- synthesize empirical research on undergraduate teaching and learning in the sciences,
- examine the extent to which this research currently influences undergraduate science instruction, and

- describe the intellectual and material resources that are required to further develop DBER.

## DEFINING DBER

The committee defined DBER as a collection of related research fields. DBER scholars in physics, chemistry, engineering, biology, the geosciences, and astronomy study similar problems, use similar methods, and draw on similar theories. However, the DBER fields also exhibit important differences that reflect differences in their parent disciplines and their histories of development.

As defined by the committee, the goals of DBER are to

- understand how people learn the concepts, practices, and ways of thinking of science and engineering;
- understand the nature and development of expertise in a discipline;
- help identify and measure appropriate learning objectives and instructional approaches that advance students toward those objectives;
- contribute to the knowledge base in a way that can guide the translation of DBER findings to classroom practice; and
- identify approaches to make science and engineering education broad and inclusive.

To address these goals, DBER scholars conduct a wide range of studies that includes basic and applied research. Both types of research are valuable and important.

High-quality DBER combines expert knowledge of a science or engineering discipline, of the challenges of learning and teaching in that discipline, and of the science of learning and teaching generally. This expertise can, but need not, reside in a single DBER scholar; it also can be strategically distributed across multidisciplinary, collaborative teams.

## SYNTHESIS OF THE LITERATURE

DBER scholars have devoted considerable attention to effective instructional strategies and to students' conceptual understanding, problem solving, and use of representations. Key findings from DBER are consistent with cognitive science research and studies in K-12 education.

To gain expertise in science and engineering, students must learn the knowledge, techniques, and standards of each field. However, across the disciplines, the committee found that students have incorrect understandings about fundamental concepts, particularly those that involve very large or very small temporal and spatial scales. Moreover, as novices in a domain, students are challenged by important aspects of the domain that can seem easy or obvious to experts, such as problem solving and understanding domain-specific representations like graphs, models, and simulations. These challenges pose serious impediments to learning.

DBER clearly shows that research-based instructional strategies are more effective than traditional lecture in improving conceptual knowledge and attitudes about learning. Effective instruction involves a range of approaches, including making lectures more interactive, having students work in groups, and incorporating authentic problems and activities.

To enhance DBER's contributions to the understanding of undergraduate science and engineering education, the committee recommended the following:

- Research that explores similarities and differences among different student populations
- Longitudinal studies—including studies of the K-12/undergraduate transition—to better understand the acquisition of important concepts and factors influencing retention
- More studies that measure outcomes other than test scores and course performance, and better instruments to measure these outcomes
- Interdisciplinary studies of cross-cutting concepts and cognitive processes

## INCREASING THE USE OF DBER FINDINGS

The committee concluded that DBER and related research have not yet prompted widespread changes in teaching practice among science and engineering faculty. Different strategies are needed to more effectively translate findings from DBER into practice. These efforts are more likely to succeed if they are consistent with research on motivating adult learners, include a deliberate focus on changing faculty conceptions about teaching and learning, recognize the cultural and organizational norms of the department and institution, and work to address those norms that pose barriers to change in teaching practice.

To increase the use of DBER findings, the committee recommended that current faculty adopt evidence-based teaching practices to improve learning outcomes for undergraduate science and engineering students, with support from institutions, disciplinary departments, and professional societies. Moreover, institutions, disciplinary departments, and professional societies should work together to prepare future faculty who understand the findings of research on learning and evidence-based teaching strategies.

## ADVANCING DBER AS A FIELD OF INQUIRY

Advancing DBER requires a robust infrastructure for research that includes adequate, sustained funding for research and training; venues for peer-reviewed publication; recognition and support within professional societies; and professional conferences. To these ends, the committee recommended that science and engineering departments, professional societies, journal editors, funding agencies, and institutional leaders clarify expectations for DBER faculty positions, emphasize high-quality DBER, provide mentoring for new DBER scholars, and support venues for DBER scholars to share their research findings at meetings and in high-quality journals. For their part, DBER scholars can increase their interactions and continue drawing on related disciplines (e.g., cognitive science, educational and social psychology, organizational change, psychometrics). Finally, Ph.D. programs and postdoctoral opportunities in the individual fields of DBER can advance DBER by educating scholars who will contribute to the educational research agenda and translate that research to others.

# Section I

## *Status of Discipline-Based Education Research*

# 1

# Introduction

This report comes at a time when our nation and our species face profound challenges. Ensuring adequate food, water, energy, and mineral resources to support a growing human population competes with the need to control negative impacts such as pollution, global climate change, and loss of biodiversity. Understanding and addressing these challenges will require all the wisdom, ingenuity, and knowledge that humans can muster. These efforts will necessarily involve people with a diverse array of educational backgrounds and expertise, including scientists and engineers. Undergraduate education in science and engineering plays a crucial role in providing future generations with the knowledge and skills to address these challenges.

Undergraduate education in science and engineering in the United States serves multiple purposes, including providing all students with foundational knowledge and skills, motivating some students to complete degrees in science or engineering, and providing students who wish to pursue careers in science or engineering with the knowledge and skills required to be successful. Students who go on to have successful careers in science or engineering must be problem solvers, skilled in quantitative reasoning and modeling, effective at communication and cross-disciplinary collaboration, and cognizant of relationships between science and society (Brewer and Smith, 2011). Students who do not pursue these careers need to understand science and engineering to serve in their roles as citizens, consumers, and leaders of business and government who need to make wise science-informed decisions in their personal and professional lives. Choosing healthcare for one's children, buying a car, voting about land-use regulations, or retrofitting one's house or business to be more earthquake-resistant are but a few of the decisions that today's undergraduates may face. Their decisions on these and other issues will be based, in part, on their confidence in the methods of science and engineering and their understanding of the findings of science and engineering.

The importance of science and engineering in preparing the technical workforce and a science-literate citizenry has drawn increased attention to the quality of undergraduate science and engineering education and how it can be improved. There are persistent concerns that undergraduate science and engineering courses are not providing students with high-quality learning experiences, nor are they attracting and retaining students in science and engineering fields (President's Council of Advisors on Science and Technology, 2012). Colleges and universities also face the challenge of serving an increasingly socially, economically, and ethnically diverse undergraduate population entering college classrooms directly from high school, after a military career or other life experiences, or from postsecondary educational experiences at another institution. Sustained attention to motivating, engaging and supporting the learning of all students who enter college science and engineering classrooms is an imperative.

Completion rates for all undergraduate students, including whites and Asians, are significantly lower in science, technology, engineering, and mathematics than in other disciplines. For example, Hispanic and African American students are as likely as white and Asian students to start college with an interest in science and engineering, but less likely to persist (National Academy of Sciences, National Academy of Engineering, and Institute of Medicine, 2011). Specifically, underrepresented racial and ethnic groups comprised roughly 30 percent of the national population in 2006, but only 9 percent of the college-educated science and engineering workforce (National Academy of Sciences, National Academy of Engineering, and Institute of Medicine, 2011).

Recognizing these challenges and the need for improvements in undergraduate science and engineering instruction, many institutions are working to identify effective approaches (Association of American Universities, 2011). Faculty members—alone or in collaboration with others—also are engaged in efforts to improve instruction, measure the efficacy of these teaching practices, and understand how students learn the concepts and practices that are fundamental to their disciplines (National Research Council, 2012; Project Kaleidoscope, 2011a, 2011b). Discipline-based education research (DBER)—by systematically investigating learning and teaching in science and engineering and providing a robust evidence base on which to base practice—is playing a critical role in these efforts.

## DEFINING DISCIPLINE-BASED EDUCATION RESEARCH

DBER is grounded in the science and engineering disciplines and addresses questions of teaching and learning within those disciplines. The roots of this type of research can be traced to the early 1900s, but DBER emerged more prominently in the 1980s and 1990s (see Chapter 2 for a detailed discussion of the history). DBER can be defined both by the focus of the research and by the researchers who conduct it. In the following sections, we define DBER and who conducts it. This definition guided the committee in identifying the relevant bodies of research, and examining how to advance DBER and strengthen its impact.

DBER investigates learning and teaching in a discipline using a range of methods with deep grounding in the discipline's priorities, worldview, knowledge, and practices. It is informed by and complementary to more general research on human learning and cognition. Although the focus of this report is learning and teaching in undergraduate institutions, DBER scholars have also examined learning and teaching in the K-12 context, particularly at the high school level.

The long-term goals of DBER are to

- understand how people learn the concepts, practices, and ways of thinking of science and engineering;
- understand the nature and development of expertise in a discipline;
- help to identify and measure appropriate learning objectives and instructional approaches that advance students toward those objectives;
- contribute to the knowledge base in a way that can guide the translation of DBER findings to classroom practice; and
- identify approaches to make science and engineering education broad and inclusive.

Thus the research has the practical goal of improving science and engineering education for all students.

Achieving these goals requires that DBER studies be grounded in expert knowledge of the discipline and the challenges for learning, teaching, and professional thinking within that discipline. All fields of DBER share a common focus on issues that are important for understanding and fostering student learning of the most crucial topics, techniques, procedures, and ways of knowing that define the particular discipline. This focus includes investigating student learning within that discipline *per se*, along with issues affecting enrollment and retention of students in classes and the adoption of best practices by instructors.

To progress toward these goals, DBER relies on several types of knowledge from outside the science or engineering disciplines: (1) the nature of human thinking and learning as they relate to the discipline of interest, (2) factors that affect student motivation to initially engage in and then to persist in the learning necessary to understand the discipline and apply findings of the discipline, and (3) research methods appropriate for investigating human thinking, motivation, and learning. By its very nature, DBER is an interdisciplinary field of study. This means that discipline-based education researchers must bridge the gaps in language, background, and ways of thinking between their home discipline and several areas of research on learning and teaching.

DBER embraces the full spectrum of research approaches for understanding human learning, cognition, and affect. Its research methods are drawn not only from the home discipline (e.g., chemistry or engineering) but also from a variety of other fields such as experimental and social psychology, education, and anthropology. Discipline-based education researchers use experimental, correlational, ethnographic, and exploratory designs, and to collect quantitative and qualitative evidence.

As with other areas of research, DBER includes a range of studies from fundamental to applied, and from theoretical to empirical. A useful framework for thinking about the range of questions that can be addressed in research was proposed by David Stokes in his book *Pasteur's Quadrant* (Stokes, 1997; see Figure 1-1). Using this framework, some DBER studies might be categorized as pure basic research, driven by a quest for fundamental understanding that is connected to the practical goal of improved education, but with no immediate application. Basic research in DBER might include research on the cognitive underpinnings of groups of students' misconceptions (see Chapter 4). Pure applied DBER, on the other hand, might include studies of the effectiveness of collaborative problem solving or the use of technology for improving classroom instruction (see Chapter 6). Many DBER studies fall in the "use-inspired basic research" category of Pasteur's quadrant. For example, researchers have investigated students' competence at authentic tasks within the discipline, such as translating between different representations of a molecule in chemistry (Cooper et al., 2010) or using diagrammatic representations to reason about evolutionary relationships among taxa in biology (Novick and Catley, in press).

DBER has sometimes been characterized by the training and professional positions of the contributing scholars rather than solely in terms of substantive focus. As discussed in Chapter 2, DBER scholars have a diverse array of backgrounds. A number of DBER scholars have a Ph.D. in a science or engineering discipline and additional training or experience in education research. Many of these scholars hold positions in natural science departments. Yet other scholars also contribute to DBER through collaborations

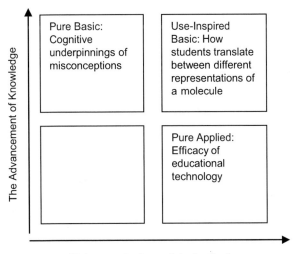

FIGURE 1-1 Pasteur's quadrant showing basic and applied DBER.

that bring together individuals with expert knowledge in science or engineering and those with expertise in education research or research on learning and teaching. The discussions in Chapter 2 of this report focus more heavily on individual scholars who have dual training in the natural sciences or engineering and experience or training in education research.

### Relation of DBER to Other Research Areas

Another way to define DBER is to consider it in relation to other fields that study learning and teaching (Bodner, 2011). In this section we consider DBER alongside three related fields: the scholarship of teaching and learning, educational psychology, and cognitive science. The category of DBER overlaps each of these other categories, but stands distinct from all of them. As noted, DBER is distinguished by an empirical approach to investigating learning and teaching that is informed by an expert understanding of disciplinary knowledge and practice. In making these distinctions we focus on the research perspective of scholars rather than on their organizational location within the institution.

#### Scholarship of Teaching and Learning

The activities that have come to be known as the scholarship of teaching and learning (SoTL) have developed in parallel with DBER. SoTL emerged from *Scholarship Reconsidered* (Boyer, 1990), which emphasized the need for classroom research and sparked conversations about teaching among colleagues at the university level. In 1997, the Carnegie Foundation for the Advancement of Teaching established the Carnegie Academy for the Scholarship of Teaching and Learning (CASTL). CASTL supported faculty fellows in developing classroom research skills, and several reports have been published on the positive impact this work has had on the fellows and their institutions (Hatch, 2005; Huber and Hutchings, 2005; Hutchings, Huber, and Ciccone, 2011). SoTL has focused on engaging faculty across disciplinary boundaries, including the humanities, social sciences, and natural sciences with their wide-ranging epistemologies and standards of evidence. SoTL emphasizes developing reflective practice and using classroom-based evidence. Some faculty engage in SoTL to inform their own work in the classroom, and some have gone on to become deeply engaged in more general education research. Thus, the boundaries between SoTL and DBER are blurred and some researchers belong to both the SoTL and DBER communities.

While DBER scholars gravitate to discipline-specific journals, SoTL researchers mostly publish in broad journals on teaching and learning such as the *Journal of College Student Development* or through the *International Journal for the Scholarship of Teaching and Learning* (*IJSoTL*). *IJSoTL* states that "SoTL is a key way to improve teaching effectiveness, student learning outcomes, and the continuous transformation of academic cultures and communities. ...[C]ollege and university teaching is seen as a serious intellectual activity that can be evidence and outcome based."[1]

### Educational Psychology Research

Educational psychologists investigate learning in students of all ages, including undergraduates (e.g., Mayer, 2011). In contrast to DBER, research in educational psychology typically focuses on general principles of learning, and the content domain under investigation is often secondary. For example, science content might be used in a study of how students learn from diagrams, but the particular science content *per se* is not necessarily the object of investigation. DBER, on the other hand, is concerned with undergraduate students' learning of a particular aspect of the scientific discipline. As a consequence, the science content that is incorporated in research in educational psychology is often not equivalent in depth or breadth to that considered in DBER.

---

[1] This statement appears on the *IJSoTL* website: http://academics.georgiasouthern.edu/ijsotl/index.htm [accessed March 30, 2012].

## Cognitive Science Research

Cognitive science is a multidisciplinary field dedicated to understanding the nature of the human mind and other intelligent systems, primarily from a basic research perspective. Theoretical investigations generally focus on issues of knowledge representation, cognitive processes, and (in humans) brain theory (Friedenberg and Silverman, 2006). Computational modeling of human thinking is a strong focus in the field. Cognitive science researchers value both novel experimental tasks (e.g., abstract puzzles) and those drawn from or modeled on real-world tasks (e.g., authentic science materials such as physics problems) (Friedenberg and Silverman, 2006).

The core disciplines of cognitive science are artificial intelligence, linguistics, anthropology, psychology, neuroscience, philosophy, and education. *Cognitive Science*, the flagship journal of the Cognitive Science Society, publishes research on intelligent systems that is multidisciplinary across two or more of these named disciplines. The journal historically has not published many articles related to education and would consider many DBER studies—such as solving problems from undergraduate physics and reasoning from diagrams in evolutionary biology—as belonging to the field of cognitive psychology, which includes the study of problem solving and diagrammatic reasoning in undergraduates. Although such studies are interdisciplinary between cognitive psychology and another science discipline, the second discipline is not considered to be part of cognitive science.

It can be difficult to determine whether studies of complex, high-level cognitive tasks such as problem solving and diagrammatic reasoning, which are related to DBER studies, belong to cognitive psychology or cognitive science. In many cases, both fields would reasonably lay claim to the research. In this report, we have opted to classify this research as cognitive science because (a) it generally is directed to and is cited by a multidisciplinary audience and (b) at the time of this report such research is much more likely to be presented at the annual meeting of the Cognitive Science Society than at the annual meeting of the Psychonomic Society (the home of cognitive psychology). In the minority of cases in which we refer to supporting research as coming from cognitive psychology, it is because that research is directed toward a cognitive psychology audience.

## Educational Evaluation

Finally, evaluation of educational interventions and programs is related to the work in DBER that measures the effectiveness of particular instructional strategies, course structures, or programs of study. Many educational evaluators use sophisticated methods for studying the implementation and impact of interventions in context, as well as efforts to take those interventions to scale. These methods include large-scale, mixed methods designs and a wide variety of quasi-experimental designs (see, for example, the journal *Educational Evaluation and Policy Analysis*). In contrast to DBER, however, these studies do not typically closely examine the nature of the science or engineering discipline being learned.

## OVERVIEW OF THE STUDY

Increased calls to improve instructional practices in the natural sciences intersect with growing interest in DBER as an important area of scholarship, generating new opportunities to apply this research. Recognizing this important juncture, the National Science Foundation (NSF) requested that the National Research Council (NRC) convene a committee to conduct a synthesis study on the status, contributions, and future directions of DBER across undergraduate physics, biology, the geosciences, and chemistry. In response to this request, the NRC convened the 15-member Committee on the Status, Contributions, and Future Directions of Discipline-Based Education Research to answer questions that are essential to advancing DBER and broadening its impact on science teaching and learning at the undergraduate level. Over a 13-month period in 2010-2011, the committee explored those questions. This report synthesizes the committee's findings.

### Charge to the Committee

The three broad elements of the committee's charge were to

1. synthesize empirical research on undergraduate teaching and learning in the sciences,
2. examine the extent to which this research currently influences undergraduate science instruction, and
3. describe the intellectual and material resources that are required to further develop DBER.

More specifically, the committee was charged with addressing the following questions:

1. What is the state of DBER scholarship as a whole and what currently is being done across each of the natural sciences? Are there research synergies across disciplines?
2. What findings are robust across disciplines?
3. What discipline-specific instructional practices are most clearly linked to increased performance across student groups (especially low socioeconomic status, minority, and female students)?

4. To what extent and how has DBER informed teaching and learning in the various disciplines?
5. What factors are influencing differences in the state of research and its impact in the various disciplines?
6. What are the resources, incentives, and conditions needed to advance this research?
7. What resources and incentives are needed to ensure that teaching and learning in the various science disciplines is informed by DBER?
8. What questions should DBER scholars prioritize in the next generation of research?

### Scope of the Study

The original charge to the committee specified that the committee consider undergraduate physics, biological sciences, geosciences, and chemistry. As work on the study began, two changes were made to the disciplines that were included. First, engineering was included because early discussions suggested engineering education research was robust and the engineering education research community was establishing an infrastructure for research. Second, a consideration of the literature in physics education research revealed that astronomy education research differed in terms of timeline and trajectory and merited inclusion as a discipline separate from physics. Thus, although astronomy typically is linked with physics at the undergraduate level, the committee decided to treat them separately for this study because astronomy education research and physics education research are at different points in their development, emphasize different methodological approaches, and involve distinct (though overlapping) student populations.

It is important to note that DBER can be a field of study within any academic discipline, in the sciences and beyond. However, because this study focuses on education research in a select set of science and engineering disciplines, throughout this report we use the term DBER to refer only to these disciplines.

In addressing the specific questions in the charge, the committee agreed on the following approaches. In question 1, we interpreted "scholarship" to encompass the community of DBER scholars in the sciences and engineering and the body of literature that those researchers generate. Determining the state of scholarship includes examining the types of questions that DBER scholars ask or the problems they study, how they study them, what counts as evidence, and key findings. It also includes examining degree programs, postdoctoral and faculty positions, conferences, professional societies, journals, and other indicators that reflect the development and identity of DBER in a discipline and its practitioners within the academic culture. And finally, determining the state of scholarship also includes a historical sense of how the research community has developed over time.

In discussing question 2, we determined that it would be necessary to summarize the findings within each discipline before analyzing the findings across disciplines. Concerning question 3, we also agreed that, depending on how the research was disaggregated, it would be useful to consider student characteristics other than socioeconomic status, minority status, and gender. However, overall, our synthesis revealed that relatively little DBER has been designed to examine group differences.

Finally, noting that question 7 creates a sense of direction for the field, we focused on ensuring the widespread use of research-based practices. We also recognized the importance of identifying whom the resources should target, and of considering future science and engineering faculty and future DBER scholars—including graduate students—when answering this question.

## Approach and Sources of Evidence

The committee carried out its charge through an iterative process of gathering information, deliberating, identifying gaps and questions, gathering further information to fill these gaps, and holding further discussions. In the search for relevant information we held four public fact-finding meetings, reviewed published reports and unpublished research, and commissioned experts to prepare and present papers. During a fifth, private meeting, we intensely analyzed and discussed our findings and conclusions.

Our approach began with an examination of the research on teaching and learning at the undergraduate level in each discipline of the study charge, with a focus on DBER. To this end, we commissioned literature reviews of DBER in astronomy, biology, chemistry, engineering, the geosciences, and physics. Equipped with this foundational understanding, we addressed issues that cut across disciplines by considering some general principles of teaching and learning from cognitive science and educational psychology. Finally, we examined a broader set of factors that influence faculty, departmental, and institutional change, and considered a set of strategies designed to promote research-based instructional and institutional change in undergraduate science instruction.

We found a limited number of studies that identify the extent to which DBER has informed instruction (study question 4). Thus, to address this question we also sought to identify the factors that influence faculty decisions about instruction, primarily through a commissioned paper that drew on the broader research about individual and institutional transformation in postsecondary education.

When considering the issues of advancing DBER (study question 6), we found that the research base was similarly sparse. Therefore, to address this issue, we commissioned papers examining the history of DBER in each of the disciplines in the study charge. Those papers helped to define DBER, designate milestones associated with the development of emerging fields, and identify relevant journals for education research in each discipline. They also enabled a cross-cutting comparison of the development of DBER. Another aspect of advancing DBER relates to preparing and placing future faculty members. However, no systemically collected data existed on graduate or postdoctoral programs or career pathways for discipline-based education researchers, so we also commissioned a paper that would allow us to explore the role of postdoctoral programs in preparing DBER faculty.

Although the committee considered information from a variety of sources during the course of this study, the conclusions we have drawn about the research on teaching and learning within each discipline give the most weight to research published in peer reviewed journals and books. Following an earlier National Research Council report (2002), we adopted the view that "A wide variety of legitimate scientific designs are available for education research. They range from randomized experiments ... to in-depth ethnographic case studies ... to neurocognitive investigations ... using emission tomography brain imaging" (p. 6). Reflecting this view, we developed a set of categories to characterize the strength of the conclusions we could draw from the available evidence (see Box 1-1).

## FOCUS AND ORGANIZATION OF THIS REPORT

The bulk of this report (Chapters 4 through 7) is dedicated to a synthesis of research on undergraduate teaching and learning in physics, chemistry, engineering, biology, the geosciences, and astronomy. With the synthesis we have attempted to strike a balance between preserving characteristics or challenges that are tied to just one or two disciplines (e.g., students' difficulties understanding deep time or that matter is made of discrete particles), and identifying general themes in science and engineering learning that cut across most disciplines (e.g., students' difficulties solving problems and interpreting visual and mathematical representations).

The report also discusses the emergence and current state of the individual fields of DBER (Chapter 2); analyzes the use of DBER findings among faculty members (Chapter 8); and provides a roadmap for the future of DBER by proposing a research agenda and identifying actions that

---

**BOX 1-1**
**Characterizing the Strength of Conclusions**
**Supported by the Evidence Base**

A Limited Level of Evidence
- Few peer-reviewed studies of limited scope with some convergence of findings or convergence with nonpeer-reviewed literature or with practitioner wisdom.

A Moderate Level of Evidence
- A well-designed study of appropriate scope that has been replicated by at least one other similar study. Often such evidence will include both quantitative and qualitative data OR
- A few large-scale studies (e.g., across multiple courses, departments, or institutions) with similar results OR
- A moderate number of smaller-scale studies (e.g., in a single course or section) with general convergence but possibly with contradictory results. If the results are contradictory, more weight might be given to studies that reflect methodological advances or a more current understanding of teaching and learning, or are conducted in more modern learning environments.

Strong Evidence
- Numerous well-designed qualitative and/or quantitative studies, with high convergence of findings.

---

postsecondary institutions, disciplinary departments, journal editors, professional societies, and funding agencies can take to support and advance DBER (Chapter 9). Because the fields of DBER have been and will continue to be an important way of improving science and engineering education, our hope is that the findings and recommendations in this report invite and assist postsecondary institutions to increase interest and research activity in DBER.

# 2

# The Emergence and Current State of Discipline-Based Education Research

Building on the definition of discipline-based education research (DBER) in Chapter 1, the first section of this chapter traces the development of DBER within physics, chemistry, engineering, biology, the geosciences, and astronomy. In that discussion, we describe the emergence and current status of DBER within each discipline, including who conducts DBER and the pathways to developing expertise as a discipline-based education researcher. These histories draw heavily on papers commissioned for this study (Bailey, 2011; Bodner, 2011; Cummings, 2011; DeHaan, 2011; Lohmann and Froyd, 2011). Next, looking across the fields of DBER, we analyze the current status of DBER overall. This analysis was guided by Fensham's (2004) criteria for characterizing the emergence of new disciplines.

## THE EMERGENCE OF DISCIPLINE-BASED EDUCATION RESEARCH

Although the trajectories of DBER across the different disciplines are distinct, they share some milestones that reflect developments within the larger context of science and education. In the late 1800s and early 1900s, concerns about the quality of learning and teaching science at the postsecondary level began to emerge, marking the first steps toward DBER. These concerns coincided with the expansion of colleges and universities in the United States (Rudolph, 1990). At this time, the focus was on the quality of science education based on the judgment of disciplinary experts rather than on a research program to improve that quality. The next common milestone occurred in the 1950s and 1960s, when the launch of Sputnik sparked a

*19*

realization that having a sufficient number of scientists and engineers in the United States was essential to remain competitive on the world stage. As part of the response, the National Science Foundation funded science curriculum development projects and involved scientists from the disciplines in that work (Cummings, 2011; Rudolph, 2002). Finally, from the 1970s through the 1990s, scholarly research that might be considered true DBER emerged and the individual fields of DBER gained recognition as fields of study within the science disciplines. Recognition of DBER can be seen in statements by professional societies, the establishment of journals and the emergence of graduate and postdoctoral opportunities.

In the following sections, we trace the development of DBER in each of the parent disciplines. The six fields of DBER are discussed in roughly chronological order, from the parent discipline where DBER, in its modern form, first emerged to those where it emerged later. We adopt this approach because fields that have developed more recently have built on the experiences of older fields.

### The Emergence of Physics Education Research

Early roots of physics education research (PER) can be traced to concerns about the quality of physics education that emerged during the late 1800s and early 1920s. These concerns led to the establishment of the American Association of Physics Teachers (AAPT) in 1930. Since 1932, AAPT has been the primary organization supporting the improvement of physics education in the United States.

The rise of PER was probably a result of the national concern about science education in the late 1950s and 1960s, which led to the involvement of natural scientists in efforts to improve science education (Cummings, 2011; Matthews, 1994). Following Sputnik, large infusions of federal funds and the emergence of many highly respected physicists as leaders in educational reform made involvement in physics education more attractive to members of that community. For every level of the education system from early elementary school to the university, new curricular content was developed that was more consistent with contemporary physics research. Large-scale efforts to reformulate undergraduate introductory physics were centered at the University of California at Berkeley, the California Institute of Technology, and the Massachusetts Institute of Technology (Feynman, Leighton, and Sands, 1964; French, 1968; Kittel, Knight, and Ruderman, 1965). At the K-12 level, physicists worked with educators to use new pedagogy that reflected the processes of science (similar to what has recently been called inquiry or scientific practices). These efforts included the Physical Science Study Committee (Finlay, 1962), Harvard Project Physics (Holton, 2003), and the Science Curriculum Improvement Study (Karplus, 1964). Although the intellectual structure of these curricula and the national support for them seemed strong, by the 1970s these reform efforts were no longer widely used. By the 1980s, only traces of them remained at any level of the U.S. educational system (Matthews, 1994).

The first groups doing work that could be called PER began systematic research programs on student difficulties at the University of California, Berkeley and the University of Washington (Cummings, 2011). The Berkeley group was a mix of physicists, educators, and educational psychologists; the Washington group was self-contained in physics. These groups influenced some physicists to join the PER effort and some physics students to search for a professional path into the emerging field of PER.

The first PER Ph.D.s graduated in the late 1970s. By the 1980s, at least a dozen universities included PER groups, which began graduating PER Ph.D.s in the 1990s (Cummings, 2011). These first Ph.D. programs in PER followed three models, which are still in use today:

1.  a Ph.D. completely within the physics department,
2.  a Ph.D. in education with the intellectual home in physics, and
3.  an interdisciplinary degree with the intellectual home in physics.

During the 1990s, students with PER Ph.D.s and physics Ph.D.s who were crossing over to PER as postdoctoral researchers began to join the faculty of physics departments. Between 1998 and 2004, 61 faculty were hired in positions with PER as their area of research (Meltzer et al., 2004).

During the late twentieth century, the organizational wall between education and physics research became more permeable. The American Physical Society (APS) reestablished its Committee on Education in 1973 (originally created in 1920, but discontinued in 1927), added an education officer in 1986, and established a Forum on Education in 1993. In 1999, APS issued a policy statement recognizing PER as part of the research portfolio of a physics department.[1] Since that time, the APS has actively promoted the improvement of physics education through PER and the use of PER-based educational practices (Cummings, 2011).

By the 1990s, increasing numbers of scholars were attending special PER sessions at the semiannual AAPT national meetings. Since 1997, the PER community has held its own annual national meeting, typically with more 200 attendees (Cummings, 2011). By the year 2000, PER sessions were held regularly at APS national meetings. With several national and international topical meetings for PER each year, physics education researchers have clearly become a vigorous community.

---

[1] For the policy statement, see http://www.aps.org/policy/statements/99_2.cfm [accessed April 19, 2012].

Despite the growth of the field in the 1990s, prior to 2000 it was difficult to publish PER. The AAPT journal, the *American Journal of Physics*, published only articles featuring practical instructional techniques and did not emphasize the research on which they were based. A PER supplement to this journal was established that accepted a limited number of articles. In 2005, following APS' recognition of physics education as an important subdiscipline of physics, the society established a *Physical Review* journal specifically for physics education research. Currently, PER is published primarily in the two AAPT journals, *American Journal of Physics* and *The Physics Teacher*, or in the APS journal, *Physical Review Special Topics— Physics Education Research*.

Although PER has expanded in the 40 years of its existence, it is still a fledgling field (Cummings, 2011). A total of 79 PER groups, with one or more faculty members, are active at colleges and universities in the United States.[2] Only about a dozen institutions grant Ph.D.s in this field, and collectively they produce a handful of graduates each year. The appointment of a PER Ph.D. to the tenure-track faculty at a major research university is still a rare occurrence. Only a few funded postdoctoral opportunities in PER are available (Cummings, 2011). PER research funding continues to be most closely linked to curriculum development in programs such as the National Science Foundation's (NSF's) Transforming Undergraduate Education in Science, Technology, Engineering, and Mathematics (TUES) Program and, to a smaller extent, by NSF's Division of Mathematical and Physical Sciences, the Department of Education's Fund for the Improvement of Postsecondary Education, the National Aeronautics and Space Administration, and the Office of Naval Research. However, NSF is also beginning to fund PER that is not directly connected with curriculum development.

## The Emergence of Chemistry Education Research

Chemistry Education Research (CER) has origins similar to PER. Concerns about the quality of undergraduate chemistry education emerged in the 1920s, but they did not lead to the establishment of a separate professional society focused on teaching. Although the *Journal of Chemical Education* was established in 1924, research focused on chemistry education did not emerge at that time.

Paralleling physics, in the 1960s a number of curriculum development programs in chemistry were catalyzed by national concerns about the need for scientists and engineers. ChemStudy and the Chemical Bond Approach were developed in response to the perception that chemistry was taught at

---

[2]For a listing of PER programs, see http://www.compadre.org/per/programs/ [accessed April 16, 2012].

the macroscopic level mainly as the classification and preparation of numerous compounds and elements. The new curricula were primarily driven by attempts to provide students with a deeper understanding of chemical principles and atomic molecular theory.

The initial forays into what would become CER were carried out by faculty members of chemistry departments who had achieved tenure and promotion to full professor in traditional areas of chemistry research. At Purdue University, these conditions gave rise to the Division of Chemical Education in the Department of Chemistry in 1981 (Bodner, 2011). The first doctoral programs in chemistry departments that awarded Ph.D. degrees for CER arose in the 1990s at the University of Oklahoma, the University of Northern Colorado, and Purdue University. The first Ph.D. in CER was awarded in 1993, with the first postdoctoral appointment in 1994. During the same time frame, tenure was also first awarded to CER faculty in chemistry departments.

In 1994, the American Chemical Society (ACS) Division of Chemical Education established a committee on CER. As the world's largest scientific society, ACS first recognized CER in its Statement on Scholarship in 2000[3] and has formally revised and renewed that commitment every two years since that time. In 2007, ACS established the Award for Achievement in Research on the Teaching and Learning of Chemistry. ACS national meetings typically include several research symposia devoted to CER. The Biennial Conference on Chemical Education (of the ACS Division on Chemical Education) also provides a growing number of CER symposia. Beyond ACS, premier scientific organizations in chemistry recognize CER to varying degrees. In 1994, the Gordon Research Conferences established the "Innovations in College Chemistry Teaching" conference, now the "Chemistry Education Research and Practice" conference, to feature the frontiers of cutting-edge CER.

The *Journal of Chemical Education* (published by the Division of Chemical Education of the ACS) and the Royal Society's *Chemistry Education Research and Practice* are the two primary publications for CER. However, as of 2012, the weekly flagship journal of the ACS, the *Journal of the American Chemical Society*, had yet to publish any CER papers.

Today in the United States, there are 29 doctoral programs where graduate students can earn a Ph.D. in chemistry for conducting research on the teaching, learning, or assessment of chemistry.[4] Most of these programs

---

[3]The statement is available at http://portal.acs.org/portal/acs/corg/content?_nfpb=true&_pageLabel=PP_TRANSITIONMAIN&node_id=1531&use_sec=false&sec_url_var=region1&_uuid=8e5645fb-ce86-447b-92f4-78010f0a29cf [accessed April 16, 2012].

[4]For a listing of CER graduate programs, see http://www.users.muohio.edu/bretzsl/gradprograms.html [accessed April 16, 2012].

require students specializing in CER to complete coursework across the core subdisciplines of analytical, bio-, inorganic, organic, and physical chemistry. These students also typically take methodology and theory courses in statistics, sociology, curriculum, cognitive science, and educational psychology. In addition to the CER dissertation research, many programs also require CER students to conduct experimental "bench" research.[5] There are no postdoctoral programs in CER.

In terms of funding, the NSF Career Program now makes awards that include CER. In 2011, the NSF Graduate Research Fellowship Program accepted DBER proposals for the first time, and one CER award was made. Typical sources of NSF funding for CER include the Division of Undergraduate Education and the Division of Research on Learning in Formal and Informal Settings. Within the Math and Physical Sciences Directorate, the Division of Chemistry at NSF does not fund CER (although it does fund some PER), in contrast to the Engineering, Biological Sciences, and Geosciences Directorates. Some other federal and private funding programs that support bench chemistry research do not typically invite CER proposals.

## The Emergence of Engineering Education Research

Although engineering shares many teaching and learning concerns with the science disciplines—after all, engineering students take about one-third of their courses in science and math—it is markedly different from other fields. Design, problem solving and application of knowledge are fundamental to engineering. Also, unique among the disciplines represented in DBER, engineering programs are externally accredited by ABET, which, as described elsewhere in this report, strongly influences engineering education and engineering education research (EER).

Engineering education emerged as an area of interest for curriculum development and pedagogical innovation in the United States with the founding of the Society for the Promotion of Engineering Education (SPEE) in 1893 (now known as the American Society for Engineering Education) (Lohmann and Froyd, 2011). In 1910, SPEE established the first periodical related to engineering education, called the *Bulletin of the Society for the Promotion of Engineering Education*. The publication changed names several times during the century that followed. In 1993, it became the *Journal of Engineering Education* and made an explicit shift to publishing research. Although this is the only journal to exclusively focus on EER, the missions of several other engineering journals, such as *Engineering Studies, European Journal of Engineering Education, International Journal*

---

[5]For a description of the content of CER graduate programs, see http://www.users.muohio. edu/bretzsl/gradprograms.html [accessed April 16, 2012].

*of Engineering Education, Engineering Education,* and *Chemical Engineering Education* include education research (Lohmann and Froyd, 2011).

The transition of engineering education to a more scholarly field of scientific inquiry occurred nearly a century after its inception—catalyzed by NSF funding for education research and development beginning in the late 1980s, and the emergence of the outcomes-based ABET Engineering Criteria in the late 1990s. In 1996, ABET specified areas of knowledge and skill development for students by which degree-granting institutions would be judged beginning in 2001 (see Chapter 3 for a description). The ABET criteria specify a range of student learning outcomes including specific knowledge and skills as well as more general habits of mind and professional conduct (ABET, 2009).

The dialogue and decisions made in the 1990s, fueled by the increasing awareness within engineering that the intuition-based approaches of the past were not producing the engineering talent required to address society's current challenges (National Academy of Engineering, 2004; National Science Foundation, 1992), paved the way for EER to become an established field of inquiry. In 2004-2005, the NSF-funded Engineering Education Research Colloquies led to the development of a taxonomy of EER organized around "five priority research areas (Engineering Epistemologies, Engineering Learning Mechanisms, Engineering Learning Systems, Engineering Diversity and Inclusiveness, and Engineering Assessment)" that merge knowledge of disciplinary engineering and the science of learning (The Steering Committee of the National Engineering Education Research Colloquies, 2006a, 2006b).

EER has begun to emerge as an interdisciplinary field seeking its own theoretical foundations from a rich array of research traditions in the cognitive and learning sciences, education, and other DBER fields (Lohmann and Froyd, 2011). EER is increasingly featured at conferences of engineering education societies around the world. Engineering education research has a strong presence in the Educational Research and Methods Division of the American Society for Engineering Education and at their two yearly conferences. The annual Research on Engineering Education Network conference is entirely devoted to engineering education research, and the Collaboratory for Engineering Education Research[6] provides online communities and resources for EER scholars analogous to On the Cutting Edge in the geosciences (described under "The Emergence of Geoscience Education Research").

In addition, EER doctoral dissertations have increased dramatically. Between 1929 and about 1980, EER doctoral dissertations were sporadic. From 1980 to 1989, between 5 and 11 of these dissertations were published

---

[6]For more information see http: //cleerhub.org/ [accessed March 30, 2012].

per year (Strobel et al., 2008). In contrast, more than 460 Ph.D. dissertations focused on engineering education between 1990 and 2010. In 2004, Purdue University and Virginia Polytechnic Institute and State University each created a Department of Engineering Education (Lohmann and Froyd, 2011).

Lohmann and Froyd (2011) conclude that EER "has established the critical physical infrastructure, e.g., centers, departments, journals, conferences, and funding, necessary for it to now devote increasing attention to its intellectual growth, e.g., conceptual and theoretical development, research methodologies, and progression" (p. 11). Svinicki (2011) argues that future progress for EER will involve collaboration with experts in a variety of other disciplines, including psychology, education, and communication.

## The Emergence of Biology Education Research

Similar to physics, chemistry and engineering concerns about the quality of biology education emerged at the start of the twentieth century. However, little research was focused on biology education. Investigators were not well known to each other, and venues for publication were scant (DeHaan, 2011).

Beginning in the 1930s, the journal *Science Education* started to shift its emphasis from solely primary and secondary instruction to include college instruction. Several other outlets for publication on biology education were established by emerging professional societies, such as *American Biology Teacher* (established by the National Association of Biology Teachers in 1938), and the *AIBS Bulletin*, which became *BioScience* (established by the American Institute of Biological Sciences in 1951).

Contributors to early research on biology education were primarily motivated by general questions of science learning, such as the relative value of lecture and demonstrations versus laboratory instruction, conceptualization versus memorizing, and the effectiveness of collaborative versus individual competitive learning (DeHaan, 2011). However, a few science faculty, prompted in particular by concerns over the prescribed laboratory exercises that had become common by the 1930s, experimented with new instructional approaches using their college biology students as participants. After the introduction of Bloom's taxonomy of intellectual behavior (1956), which clarified the distinction between memorizing factual information and learning for understanding, biology educators sought ways to promote conceptual learning. In the 1980s and 1990s, following the lead of PER, biology educators began to document common student misconceptions in biology (e.g., Pfundt and Duit, 1988).

Much of the biology education research (BER) published before 2000 was descriptive, for example, reporting the development of a new course or laboratory module and student reactions to it. However, since the 1990s there has been a gradual shift toward more analytical and quantitative studies of teaching and learning, stimulated in part by PER and in part by entry of more "border crossers" into the field (i.e., biologists who became interested in researching the effectiveness of instruction and brought analytical approaches from their scientific work into their education research). Concept inventories modeled on the Force Concept Inventory from physics (Hestenes, Wells, and Swackhamer, 1992) were developed for several areas of biology (D'Avanzo, 2008) to help monitor the effectiveness of instruction in dispelling student misconceptions (see Chapter 4 for a discussion of concept inventories). A parallel contributing factor to this shift was the establishment of new journals demanding different standards of evidence for the value of instructional interventions.

A review of the literature from 1990-2010 commissioned for this study (Dirks, 2011) identified about 200 studies that reported data on student learning, performance, or attitudes in college biology courses. Most (83 percent) of the 200 articles reviewed by Dirks (2011) were published since 2000. These articles appeared in more than 100 different journals. However, most were published in just four journals—the *Journal of Research in Science Teaching* (*JRST*) and three journals established since 1990: *Journal of College Science Teaching* (established 1994); *Advances in Physiology Education* (established 1996); and *Cell Biology Education,* later renamed *CBE-Life Sciences Education* (established 2002). *CBE-Life Sciences Education*, an open-access online journal sponsored by the American Society for Cell Biology with support from the Howard Hughes Medical Institute, has had from its inception an editorial policy of publishing only articles that include clear evidence for the value of instructional interventions based on student assessments.

Biology is organized into a large number of subfields with many professional societies. In contrast, the BER community is emerging in a more centralized way. In 2010, the BER community established the Society for the Advancement of Biology Education Research (SABER) with the explicit goal of advancing the field of undergraduate BER. SABER provides a community for disseminating information on BER, fostering research collaborations, and promoting more uniform standards for the training of BER scholars. Data collected from participants at the first annual SABER meeting in July 2011 indicate that a defined pathway of training for students who aspire to a career in BER is emerging, but that considerable variation remains.

Demand is increasing for graduate programs that train students in both biological and education research. Currently fewer than a dozen such programs in U.S. departments of biology have at least two BER tenure-track faculty and prescribed degree requirements that generally involve the same comprehensive examinations given to biology students, but include no experience conducting disciplinary research. About 10 additional biology departments offer the opportunity for a Ph.D. in BER to the students of an individual, tenure-track BER faculty member.[7]

The small number of biology departments offering BER Ph.D.s reflects the persistent, general ambivalence toward BER as a subfield of biology. Only a handful of biology departments have recognized BER as a subfield that should be represented on the biology faculty. Except for the more senior border crossers, much of the BER in biology departments is conducted by nontenure-track researchers who do not mentor graduate students. In addition, there are limited postdoctoral education opportunities for biology Ph.D.s who wish to become education researchers or education research Ph.D.s who seek advanced training in biology.

## The Emergence of Geoscience Education Research

Paralleling physics, chemistry, and biology, concern about geoscience education first emerged in the late 1800s. With an emphasis on physical geography and meteorology, the focus of these efforts was narrower than that of modern geosciences. Moreover, as in the other disciplines, this early concern about the quality of education did not emphasize research on education.

As part of the curriculum development efforts of the 1950s and 1960s the Earth Science Curriculum Project (ESCP) was commissioned (post-Sputnik, but pre-plate tectonics). The curriculum developers strongly emphasized laboratory and field study, in which students actively participated in the process of scientific inquiry rather than repeating step-by-step exercises (Irwin, 1970). These and other emphases of the ESCP remain targets for contemporary geoscience education research: science as inquiry, the universality of change, the flow of mass and energy in the complex Earth system, the significance of Earth components and their relationship in space and time, and the comprehension of scale.

In 1996 an NSF advisory panel on geoscience education recommended that the Directorate for Geosciences and the Directorate for Education and Human Resources both "support research in geoscience education, helping geoscientists to work with colleagues in fields such as educational and cognitive psychology, in order to facilitate development of a new generation of geoscience educators" (National Science Foundation, 1997). Spurred in part by the ensuing support, geoscience education research (GER) began to coalesce as a recognized field of scholarship in the 2000s. Before that time, seminal work was being conducted by border crossers (those who conduct

---

[7]For a listing of graduate programs in BER, see https://saber-biologyeducationresearch. wikispaces.com/Graduate+Programs+in+BER [accessed April 16, 2012].

research in both GER and a traditional geoscience discipline) (Dodick and Orion, 2003; Kali and Orion, 1996; Kern and Carpenter, 1984, 1986; Orion and Hofstein, 1994; Orion et al., 1997).

A catalytic event in the history of GER was the 2002 "Wingspread" workshop, sponsored by NSF and the Johnson Foundation, on "Bringing Research on Learning to the Geosciences" (Manduca, Mogk, and Stillings, 2004). The workshop was held for the dual purposes of identifying how research from other fields could be applied to geoscience education and jump-starting a research agenda in GER. Since the Wingspread workshop (although not necessarily as a direct consequence), the number of geoscientists who engage in GER either full or part time and the rate of production of GER studies has grown.[8] Indeed, a new field of scholarship is emerging under the heading of "geocognition," which seeks to identify what it means to be an expert geoscientist, and how to facilitate the transition from novice to expert (Clary, Brzuszek, and Wandersee, 2009; Petcovic, Libarkin, and Baker, 2009).

As of 2012, fewer than a half dozen faculty nationwide had achieved promotion to tenure based on a GER portfolio, and a similarly small handful had received a Ph.D. in GER from a geoscience department. Graduate degree programs in GER are typically hosted within geoscience departments and have the same degree requirements as any other advanced degree awarded by these departments. As with BER, the balance between "geoscience" and "education" degree requirements varies from program to program. At least six universities now offer a Ph.D. in GER through their college of education, college of science, or a combination.[9] Employers of geoscience education researchers are primarily large state universities.

Geoscience education researchers find collegial support in the National Association of Geoscience Teachers (NAGT),[10] the National Earth Science Teachers Association (founded in 1983, with a focus on K-12 earth science education), the Geoscience Education Division of the Geological Society of America (GSA), and the American Geophysical Union (AGU). Professional societies, particularly GSA and AGU, have hosted an increasing number of education research sessions at their annual meetings. These meetings have exposed the larger geoscience community to emerging GER results, helped to establish GER as a respected field of scholarly work, and encouraged more colleagues to participate in GER. In 2002, GSA hosted a distinguished lecture symposium *Toward a Better Understanding of the Complicated*

---

[8]For a listing of graduate programs in geocognition and geoscience education research, see https://www.msu.edu/~libarkin/geocoggradprog.html [accessed April 16, 2012].

[9]For more information, see https://www.msu.edu/~libarkin/geocoggradprog.html [accessed April 16, 2012].

[10]The National Association of Geoscience Teachers began in 1938 as the Association of College Geology Teachers.

*Earth: Insights from Geologic Research, Education, and Cognitive Science.* Also that year, GSA published a special paper volume *Earth and Mind: How Geologists Think and Learn About Earth* with contributions from master geoscientists, learning scientists, and geoscience educators (Manduca and Mogk, 2006).

The development of venues for publication of GER has paralleled the growth in the field. NAGT launched the *Journal of Geology Education* (*JGE*) in 1951. The names of both the organization and the journal were changed from "Geology" to "Geoscience" in 1995 to reflect the broadened scope of the field. For many decades, *JGE* was primarily a journal for geoscience faculty to exchange teaching ideas and pedagogical content knowledge. In 2001, the journal began a regular column on research in education, applying ideas from other DBER fields to geoscience education. As of 2009, *JGE* has changed its guidelines for contributors and the review criteria for articles to move toward a journal that publishes a mix of GER and SoTL.

Funding for GER comes primarily from NSF. Within the Directorate for Geosciences, the Geoscience Education Program gives small grants that can be used as seed money for pilot GER projects. Some GER researchers have been funded by NSF's TUES program for applied research in the context of building educational pedagogy and materials, and from NSF's Research and Evaluation in Education in Science and Engineering (REESE) Program for more basic research.

## The Emergence of Astronomy Education Research

At the collegiate level, astronomy is closely aligned with physics, and the two fields are often part of a combined Department of Physics and Astronomy. As a result, the emergence of astronomy education research (AER) has mirrored the emergence of PER, with about a 20-year lag (Bailey, 2011). The tight connection with physics and with PER, combined with the ability to learn from PER's experience, has allowed AER to emerge relatively quickly as a separately recognized field. Astronomy education research is generally considered a part of astronomy, and most AER scholars define themselves as astronomers who study education.

Professional societies have played a critical role in the emergence and establishment of AER as a recognized discipline. Position and policy statements in support of PER by the American Physical Society and the American Association of Physics Teachers led to the development of similar statements about AER by the American Astronomical Society (AAS) in 2002.[11] Perhaps more importantly, in 2001, AAS, the Astronomical Society

---

[11]The position statement is available at http://aas.org/governance/resolutions.php#edresearch [accessed April 22, 2012].

of the Pacific (ASP), and the National Optical Astronomy Observatories established the *Astronomy Education Review*—a peer-reviewed journal that archives AER studies and applications to the teaching and learning of astronomy, broadly defined. At about the same time, AAS and ASP began issuing calls for AER papers to be presented at their meetings and began inviting AER scholars to serve as plenary speakers at their conferences. These developments notwithstanding, AER still is frequently published in PER-dominated journals and presented at PER-dominated professional conferences.

AER is funded by the same sources as PER. Research centers for AER, such as CAPER (the Center for Astronomy and Physics Education Research), and Ph.D. training programs are only just beginning to have consistent prominence and funding streams. Most contemporary AER scholars are border crossers from traditional astronomy research into education research or came from PER Ph.D. programs that include astronomy. Postdoctoral training has played a role in retraining astronomers as AER scholars. Based on the committee's knowledge of the initial placement of recent graduates of AER Ph.D. programs, most AER scholars pursue faculty positions at colleges or universities that focus on teaching rather than research.

## THE CURRENT STATE OF DBER

The brief histories presented in the previous sections reveal some of the similarities and differences across the fields that comprise DBER. This section addresses the current status of the DBER fields, with an emphasis on individuals and programs housed in disciplinary departments. Our discussion is guided by a taxonomy developed by Fensham (2004) through his analysis of the emergence of science education research.

Fensham's taxonomy includes three major categories that were adapted from the natural sciences: outcome, research, and structure. Outcome refers to the implications of research for practice. The research category emphasizes the nature of the research in the field, including both methodology and theoretical frame. Structure focuses on the research and training infrastructure of the field. For each category, Fensham outlines criteria that can be used to characterize the state of the field. For outcome he posits a single criterion, namely implications for practice. Historical analysis of DBER fields supports this outcome as the original motivation for DBER. For research, the criteria include sufficient scientific or engineering knowledge to conduct the study, distinctive questions, conceptual and theoretical development, research methodologies, progression in the research over time, model publications with clear methodologies, and seminal publications that further develop the field. As discussed in Chapters 4 through 7, DBER fields fulfill the research criteria to varying degrees.

Structural criteria include programs of training in the specialty, academic recognition, research journals, professional associations, research conferences, and research centers. Although each of the DBER fields in this report has grown to its current level of maturity along a unique trajectory, most now show evidence of meeting Fensham's structural criteria. We elaborate on the structural criteria here because they offer a useful framework for creating a more coherent picture of the current status of DBER across the fields of physics education research, chemistry education research, engineering education research, biology education research, geoscience education research, and astronomy education research.

## Academic Recognition

Fensham's (2004) definition of academic recognition is having faculty at the full professor level within the field. As noted in the previous sections, all of the DBER fields have achieved this goal.

The DBER fields vary in terms of acceptance and awareness of what faculty with DBER specialties can contribute to a department. The committee found very little published research on this phenomenon. However, a series of papers on Science Faculty with Education Specialties (SFES) of the California State University system illustrates the challenges some faculty encounter (Bush et al., 2006, 2008, 2011). In this research, SFES self-identified, and 58 percent reported being engaged in science education research. Thus this group includes some DBER scholars, but is not representative of the DBER community. The challenges that SFES face in the California State University system include access to departmental resources and demands on their time for teaching, unusually high expectations for transforming department-wide teaching, and other departmental service. DBER scholars in other settings may, or may not, face similar challenges, depending on their context.

More generally, institutions do not always recognize the distinction between education specialists whose primary focus is on teaching and DBER scholars who conduct research on teaching and learning. A review of some DBER-related academic job offerings in 2011 (CER Listserv[12]), and of other discussions concerning hiring of DBER scholars (Bauer et al., 2008) echo the theme that disciplinary departments still have diverse, competing, and sometimes imbalanced expectations for teaching, research, and service.

---

[12]See http://listserv.muohio.edu/archives/cer.html [accessed April 16, 2012].

## Research Training

Because DBER is inherently interdisciplinary, conducting DBER requires deep knowledge of the content and ways of knowing in the science or engineering discipline and expertise in conducting research about how humans think and learn. Three predominant pathways currently exist for developing expertise to conduct DBER. First, as discussed in previous sections, a growing number of science and engineering departments offer DBER Ph.D. programs. Second, limited numbers of postdoctoral DBER positions are available to provide additional training and experiences for individuals who have DBER Ph.D.s or individuals who have Ph.D.s in one of the traditional science and engineering disciplines. Lastly, border crossers whose Ph.D. and research experience are grounded in a traditional research field within a science discipline (e.g., high-energy particle physics, organic chemistry, developmental biology, or marine geology) can move into DBER through sabbatical opportunities or collaborations (Bodner, 2011). This pathway into DBER is particularly common in newer fields such as biology education research, geoscience education research, engineering education research, and astronomy education research. We discuss each of these pathways in turn.

### Formal DBER Graduate Programs

Although formal graduate programs in DBER exist and continue to emerge, they vary considerably in their organization, size, and curricular foci. We define a graduate program as an institutionally recognized program with a coherent set of standards for course requirements, comprehensive examinations, and research. Ideally a program has more than one faculty member. Given the interdisciplinary nature of DBER, some of the affiliated faculty members may be in social science departments (e.g., psychology, cognitive science), or schools of education. Using this definition, it is possible that an institution may grant Ph.D.s in a field of DBER, but not offer a formal doctoral program.

Physics education research programs typically are housed in physics departments, but often have some connection to schools of education. Successful students in both types of departments are generally awarded physics (rather than PER) degrees (Beichner, 2009). Programs in chemistry generally are located within chemistry departments, and students typically are admitted as chemistry graduate students and are awarded chemistry (not CER) degrees.[13] By contrast, engineering education programs are found in

---

[13]For a listing and description of CER graduate programs, see http://www.users.muohio. edu/bretzsl/gradprograms.html [accessed April 16, 2012].

engineering schools and in schools of education and both kinds of programs can lead to degrees in engineering education research (Lohmann and Froyd, 2011). Engineering education departments that offer graduate degrees are emerging, rather than having separate EER programs within each of the engineering specialty departments. Joint degree programs between colleges of engineering and education, such as at Ohio State University and the University of Michigan, also are gaining ground in EER (Lohmann and Froyd, 2011). A small number of BER doctoral programs are located in biology departments, and a few programs have connections to education departments. In those programs, students are awarded biology (not BER) degrees.[14] Geoscience and astronomy departments typically offer Ph.D.s through individual faculty within a disciplinary department, rather than through formal programs (Libarkin, personal communication).[15]

Graduate education in DBER is itself ripe for further study and exploration. As DBER fields mature, a growing number of researchers have been trained in DBER graduate programs and are now in academic positions. Now is the time to ask questions, not only about the outcomes of a DBER graduate education (job placement, research productivity/contributions, etc.), but also about best practices for educating graduate students in DBER. These studies would be valuable additions to the literature, and could help to guide the development of programs in newer fields such as astronomy, biology, and geoscience education. Broader guidance about supporting and evaluating interdisciplinary research and education at the undergraduate and graduate levels is documented in a National Research Council (National Academy of Sciences, National Academy of Engineering, and Institute of Medicine, 2005) report on facilitating interdisciplinary research.

One important question related to DBER graduate study is where students find employment after completing the Ph.D. doctoral graduates in physics education often take a faculty position immediately after graduation, and the postdoctoral and teaching positions often outnumber the supply of graduates (Beichner, 2009). There is little documentation of whether other DBER fields follow this trend. More work is needed to understand the trajectories of students who complete graduate study in chemistry education research, engineering education research, biology education research, geoscience education research, and astronomy education research.

---

[14]For a description of BER graduate programs, see https://saber-biologyeducationresearch. wikispaces.com/Graduate+Programs+in+BER [accessed April 22, 2012].

[15]For a listing of GER and other DBER programs, see https://www.msu.edu/~libarkin/ about_programs.html [accessed April 23, 2012].

## Postdoctoral DBER Positions

Postdoctoral education is the norm in the sciences and is increasingly prevalent in engineering. Within DBER, postdoctoral experience provides an entry point for individuals with traditional science or engineering graduate degrees to gain expertise in education research, or for individuals with DBER graduate education to develop greater sophistication in this interdisciplinary field. Although a small number of DBER postdoctoral positions arise through individual or institutional grants, no specific funding program currently exists to support DBER postdoctoral fellows. The best insights into postdoctoral positions in DBER can be gleaned from an analysis of the NSF Postdoctoral Fellowships in Science, Mathematics, Engineering, and Technology Education (PFSMETE) that supported 62 DBER postdoctoral fellows from 1997 to 1999 (Libarkin and Finkelstein, 2011).

Two-year PFSMETE fellowships were awarded to postdoctoral fellows in mathematics, physics, the geosciences, biology, chemistry, and engineering with the explicit goal of fostering boundary crossing in education research, practice and leadership. Libarkin and Finkelstein (2011) point out that ideally, PFSMETE fellows infused educational programs with their own scientific background and simultaneously infused scientific disciplines with tools from education, psychology, and other social sciences. Many fellows had not worked with colleagues in education, psychology, and cognitive science departments before beginning their fellowships. Indeed, interviews with PFSMETE fellows revealed significant rifts that still existed in 2000 between the STEM disciplines and researchers in education departments. Several fellows expressed concern that they were not welcomed by either scientists or educators. Any postdoctoral training program of this type, which seeks to connect distinct research communities, must explicitly acknowledge the divide and build an infrastructure designed to help bridge the differing worlds (Libarkin and Finkelstein, 2011).

Though short-lived, the role of PFSMETE in establishing DBER itself should not be underestimated. At a time when DBER was beginning to emerge within key disciplinary fields, the PFSMETE program provided both the imprimatur of the NSF and the human resources to staff emerging research efforts within DBER. PFSMETE's ultimate impact on science, education, and the bridges that connect these fields remains to be seen and is the subject of current studies.

In 2011, NSF introduced a new program, Fostering Interdisciplinary Research on Education (FIRE). The program is designed to bring together pairs of scholars, one with STEM expertise and one with education research expertise, in a mentoring relationship. Both participants must have graduate degrees, but unlike PFSMETE, individuals can apply at any point in their postgraduate career. Thus this program can support both postdoctoral fellows and border crossers, as discussed next.

## Border Crossers

Although formal graduate programs and postdoctoral fellowships in some fields are educating DBER scholars, many of the established DBER scholars were educated in traditional disciplinary graduate programs and migrated into DBER. Such border crossing is common as any new field develops, especially in the absence of formal programs for graduate students. The proportion of border crossers in each discipline of DBER varies, and quantifying how many self-identified discipline-based education researchers arose through this frontier path is difficult because the relevant data have not been systematically collected or compiled. A clear need exists to follow the academic trajectories of border crossers, as well as those DBER scholars whose graduate training was in a formal DBER program.

## Summary

DBER demands expertise in the discipline and in education research, which presents challenges and opportunities to designing effective pathways into the field. DBER scholars develop the needed expertise through several different pathways that often, but not always, begin at the graduate or postgraduate rather than the undergraduate level. Currently, graduate study in DBER can be pursued in a limited number of multifaculty graduate programs, and more commonly, departments with individual scholars who support graduate research in a DBER field. Postdoctoral positions provide a mechanism for further education for DBER graduates and for individuals with a traditional science background moving into DBER. Border crossing has been, and continues to be, a common mode of populating DBER fields in their early stages. Border crossers develop expertise through a range of venues, including collaborations and sabbaticals.

## Professional Associations and Research Conferences

Graduate and postdoctoral education opportunities within one department at one university are necessary, but not sufficient, to establish the identity of a new discipline. As discussed under "The Emergence of DBER," each DBER field has one or more professional organizations that support education research through policy statements, publication venues, and conference sections. Many of these professional homes are sections of larger disciplinary professional societies. For example, the American Chemical Society's Division of Chemical Education has a Chemical Education

Research Committee. Within the American Society for Engineering Education, the Educational Research and Methods Division supports EER. The numerous biology professional societies support education research to varying degrees, while SABER encompasses all fields of biology with a singular focus on BER. These disciplinary networks facilitate communication among DBER scholars within the disciplines.

However, there are very few formal ways for DBER scholars from different disciplines to interact with each other at the national level. For some DBER scholars, the National Association for Research on Science Teaching and the American Educational Research Association provide more general venues, and could be the sites of cross-disciplinary interaction.

The role of conferences on science and engineering education in supporting the growth of DBER and disseminating DBER findings has not been the subject of a formal research study. Yet these events (sometimes sponsored through disciplinary societies, sometimes as independent initiatives funded through federal grants) are potentially important ways to attract scholars into DBER and provide a venue for DBER scholars to engage with peers.

## Research Journals

Although conferences to present recent research findings are plentiful and readily accessible to DBER scholars, journal publication can present challenges for some discipline-based education researchers. Some tension exists between publication venues that are intended to share research findings among researchers and venues that are intended to inform instructors of the findings of DBER that might be useful in their classrooms. Publications intended for practitioners to support change in classroom teaching generally earn less professional recognition than research-focused journals and may have lower standards for the rigor of the research. High-quality research papers published in journals that practitioners are less likely to read may have less influence on classroom culture. The tension is unavoidable in fields that cover the spectrum from applied to basic research on the learning and teaching of undergraduate science and engineering.

Journal impact factors—or how frequently a journal has been cited in a given period of time—provide another perspective on the current state of DBER journals. However, they are influenced by the tension noted above. Most DBER-specific journals (e.g., *American Journal of Physics, Chemistry Education Research and Practice, Journal of Biological Education, Journal of Chemical Education*) have less status, with impact factors below 0.80, or in the case of the *Journal of Geoscience Education*, are not included in impact factor indices. The *Journal of Engineering Education, CBE-Life Sciences Education*, and *Physical Review Special Topics-Physics Education*

*Research* are exceptions; in these journals, the editorial policy is tipped toward the researcher as opposed to the instructor who uses DBER findings in the classroom. A number of general education and science education journals that are potential venues for publishing DBER papers (e.g., *American Educational Research Journal, Journal of Research in Science Teaching, Journal of the Learning Sciences, Learning and Instruction, Cognition and Instruction, Review of Educational Research, Science Education*) have considerably higher impact factors (1.6-2.4).

Discipline-based education researchers might encounter close scrutiny regarding the prestige of their field's journals. Faculty who are not yet tenured may question the merit of submitting manuscripts to journals with impact factors significantly lower than those in which their disciplinary peers are publishing. Potential consumers or evaluators of the research may conclude that the results from studies published in such journals are not of high quality. The fact that there is an education research journal that is part of the highly respected *Physical Review* series has been seen as an important advancement of the field of PER. Likewise, the decision at *Science,* which has an impact factor above 30, to publish papers on education is a significant advancement for all DBER fields.

## Research Centers

The last of Fensham's (2004) structural criteria is research centers, defined as nuclei of established scholars with funding to specifically support their research. In Fensham's view, research centers are important because they provide the intellectual community to advance research and training in the field. The inherent tension in DBER between advancing research and applying research findings to improve education is reflected in the status of research centers.

The most common situation across the fields of DBER is for a disciplinary department to have one discipline-based education researcher on the faculty. Within a few universities, organizational structures or centers support DBER scholars and foster DBER scholarship, for example, the Center for Research on College Science Teaching at Michigan State University and the Center for Research and Engagement in Science and Mathematics Education at Purdue University.

While centers devoted exclusively to DBER are rare, education centers (centers for learning and teaching) focused on improving undergraduate education have provided sites for faculty, postdoctoral fellows, and graduate students to engage with DBER and related research. For example, in engineering education, the first NSF-funded "center" for engineering research with a focus on engineering education was established in 1999. Since that time more than a dozen additional centers have been established around the country, coordinated through the National Academy of Engineering-supported Center for Advancement of Scholarship on Engineering Education (Lohman and Froyd, 2011). These centers provide some support and community for scholars migrating into engineering education, and strongly support effective teaching.

While most centers are located at large universities, the Science Education Resource Center (SERC) located at Carleton College is unique in serving as a physical and a virtual center, connecting colleges and universities across the country where geoscience is taught (Manduca et al., 2010).[16] With a hybrid mission to support the use of education research in practice and to engage in DBER, SERC provides educational experiences for postdoctoral fellows and border crossers. It also hosts the InTeGrate STEP center, which is developing research-based undergraduate curricula in the geosciences and studying student learning on the campuses of multiple partners.

Centers supporting the scholarship of teaching and learning on university campuses could help to create a pathway for the migration of faculty and graduate students into DBER. Examples include institutions in the NSF-funded Center for the Integration of Research, Teaching, and Learning (CIRTL) network.[17] On these campuses, CIRTL engages graduate and postdoctoral students and faculty members in the exploration of best practices for undergraduate teaching and learning in science and engineering. The Science Education Initiatives at the University of Colorado, Boulder, and the University of British Columbia[18] promote collaborative DBER across five science departments in a variety of ways, including through the education of postdoctoral science teaching fellows. These fellows come from a variety of disciplinary backgrounds but share an interest in making DBER at least a part of their future careers (see Chapter 8 for more details).

## Collaborations

Although most DBER is housed within single academic departments, DBER is also conducted by interdisciplinary teams. It can take a considerable amount of time and effort for interdisciplinary teams with professional expertise across several disciplines (e.g., chemistry, biology, computer science, and cognitive science) to establish common ground and become productive, but such teams can be instrumental in attacking some of the larger problems in human learning faced by the science disciplines. Indeed, several of the commissioned papers for this committee's work noted the

---

[16]For more information, see http://serccarleton.edu [accessed April 21, 2012].
[17]For more information, see http://www.cirtl.net/mission [accessed April 16, 2012].
[18]For more information, see http://www.colorado.edu/sei/ [accessed April 16, 2012].

importance of interdisciplinary collaborations for advancing their field of DBER (Docktor and Mestre, 2011; Svinicki, 2011).

Interdisciplinary teams are currently active in most of the DBER fields studied in this report. At Clemson University, for example, a department of engineering and science education has been established to facilitate interdisciplinary work; that department offers a Ph.D. in engineering and science education. As another example, geoscientists identified spatial reasoning as a key area of research for geoscience education research and initiated collaborative work with cognitive scientists (Kastens and Ishikawa, 2006; Manduca, Mogk, and Stillings, 2004; National Research Council, 2006). Collaborations across DBER fields, however, are less common. One example is a joint effort between the geoscientists at Carleton College's Science Education Research Center and a group of biology education researchers working on genomics education curriculum development and research.[19] Additionally, the Colorado Learning Attitudes about Science Survey (CLASS)—initially developed to measure novice-to-expert-like perceptions in physics learners—was adapted for chemistry and biology learners as a result of collaborations among DBER groups at the University of Colorado and the University of British Columbia (Semsar et al., 2011).

Pathways to establish interdisciplinary research are not straightforward (National Academy of Sciences, National Academy of Engineering, and Institute of Medicine, 2005). A few NSF programs (Research Initiation Grants in Engineering Education, REESE, FIRE) offer funding to promote the development of such teams, but interdisciplinary research is risky. Tenure and promotion committees may not take into account the time and energy necessary to become acculturated into a new field. This situation poses particular challenges for nontenured faculty in DBER to engage in interdisciplinary research (Rhoten and Parker, 2004). The National Academy of Sciences, National Academy of Engineering, and Institute of Medicine (2005) identify some of these challenges and discuss changes needed in the policies that govern hiring, promotion, tenure, and resource allocation to facilitate successful interdisciplinary collaboration.

## Summary

These brief histories of the fields that comprise DBER, together with the analysis using Fensham's structural criteria, offer insights into how DBER has developed. Each DBER field is anchored within the parent discipline but varies in the extent to which it is recognized as a fully fledged subdiscipline. In addition, the fields of DBER have engaged in only limited interaction with each other. As a result, DBER as a whole is an area of

---

[19]For more information, see http://serc.carleton.edu/genomics/ [accessed April 12, 2012].

study, but at this point cannot lay claim to being a field in the way that the individual DBER fields can. The newer DBER fields are emerging in a more purposeful way by leveraging prior work in physics education research and, particularly in the case of engineering education research and geoscience education research, working collaboratively with cognitive scientists and other social scientists.

Although multiple pathways to becoming a DBER scholar are, and will likely continue to be the norm, careful attention to what constitutes quality education in DBER at the graduate and postdoctoral levels is needed because professional standards of preparation within communities are nascent. There is almost no research tracking the success of DBER graduates, and none at all relating the professional success of DBER scholars to the nature of their backgrounds and preparation. To date, little attention has been directed on preparing undergraduate students for DBER careers or even making undergraduates aware that DBER exists. Sufficient, rigorous preparation in the science or engineering discipline *and* education research presents a challenge.

Professional societies have a role to play in both establishing and disseminating professional standards, as is happening in physics education research and chemistry education research. Biology education research faces the particular challenge of communicating with more than 100 professional biological research societies within the United States. The formation of SABER, which cuts across the biological subfields, should attenuate this disparity with its singular focus on education research.

Funding for research and training is uneven across the fields of DBER. DBER scholars receive funding from a mix of sources: those that are dedicated to research in the parent discipline and those that are dedicated to research on teaching and learning more broadly. The relative proportion of funding from each of these sources varies across the fields of DBER.

The analysis using two of Fensham's structural criteria—journals and research centers—reflects the predictable tension in DBER between advancing the research itself and increasing the use of DBER findings. Education research centers, funding programs, and some journals blend both goals. As in any discipline, DBER scholars strive for high quality research, which will be evaluated more fully in subsequent chapters of this report. Many DBER scholars, their disciplinary colleagues, their professional societies, and funding agencies are motivated by the critical need to reform science and engineering education informed by DBER findings. Clearly articulating the distinction between discipline-based education research and the application of DBER findings—and embracing the value of both—is important for ensuring continued advancement of the research, promoting improvement in undergraduate education, and enhancing synergies between the efforts.

## SUMMARY OF KEY FINDINGS

- *Discipline-based education research (DBER) is a small but growing field of inquiry. At this time, most efforts to develop and advance DBER as a whole are taking place at the level of the individual fields of DBER.*

- *Across the disciplines in this study, DBER is in different stages of development. DBER scholars and the individual fields of DBER have made notable inroads in terms of establishing their fields but still face challenges in doing so.*

- *DBER is inherently interdisciplinary, and the blending of a scientific or engineering discipline with education research poses unique professional challenges for DBER scholars.*

- *There are many pathways to becoming a discipline-based education researcher. At the time of this study, many established DBER scholars were trained in traditional disciplinary graduate programs and migrated into DBER. These border crossers are particularly common in the fields of biology education research, geoscience education research, and astronomy education research.*

- *Conducting DBER and using DBER findings are distinct but interdependent pursuits.*

- *Education research centers enable faculty to use DBER findings, introduce students to DBER as a career option, and support collaborations among faculty. Few of these centers currently exist, and even fewer have a singular focus on DBER.*

# Section II

## *Contributions of Discipline-Based Education Research*

# 3

# Overview of Discipline-Based Education Research

As the previous chapters show, discipline-based education research (DBER) is a relatively new area of research composed of a set of loosely affiliated fields with common goals and methods. The fields share some common history, but follow unique trajectories that reflect the characteristics of their parent disciplines. In addition, DBER has close ties to related research on teaching and learning in education and psychology.

In this chapter, we provide an overview of the research foci of the fields of DBER and consider their similarities and differences. This overview sets the context for the more detailed synthesis of DBER presented in Chapters 4 through 7. In the following sections we discuss the substantive focus of research in each field of DBER, typical methods used across the fields of DBER, and the relationship of DBER to broader principles and theories of learning and instruction. The chapter concludes by identifying some key strengths and limitations of DBER as a whole.

## SCOPE AND FOCUS

Across the fields of DBER, broad-level learning goals drive instruction and the concomitant research on instruction. The different disciplines of science and engineering continue to clarify goals regarding core ideas, crosscutting concepts, and science and engineering practices. Participants in a 2008 workshop series on promising practices in undergraduate science, technology, engineering, and mathematics education identified the following general learning goals for students, which also are relevant for DBER (National Research Council, 2011):

- Master a few major concepts well and indepth
- Retain what is learned over the long term
- Build a mental framework that serves as a foundation for future learning
- Develop visualization competence, including the ability to critique, interpret, construct, and connect with physical systems
- Develop skills (analytic and critical judgment) needed to use scientific information to make informed decisions
- Understand the nature of science
- Find satisfaction in engaging in real-world issues that require knowledge of science

The committee acknowledged the difficulty of identifying a common set of learning goals for science education at the undergraduate level because the missions and goals of courses and programs vary widely. Thus, this list does not represent our consensus on learning goals for undergraduate science education. However, as the following discussions of scope reveal, these goals are reflected to some extent across the fields of DBER.

## Physics Education Research

The extensive scope of contemporary physics education research has been reviewed by Docktor and Mestre (2011). Over time, the focus of inquiry has expanded from narrow investigations of students' difficulties in learning specific concepts to reflect the realization that improving physics learning is a complex and multifaceted problem. As a result of this shift, current physics education research addresses the following topics:

- characterizing students with respect to conceptual knowledge, problem solving, use of representations, attitudes toward physics and toward learning more broadly, knowledge of scientific processes, and knowledge transfer;
- defining goals for physics instruction based on rates of student learning, needs for future learning, transfer, or population diversity;
- developing curricular materials and pedagogies to facilitate conceptual change, improve problem-solving skills and the use of representations, improve attitudes toward physics and general learning, or provide experiences with the practices of science;
- investigating how students and instructors use curricular materials and pedagogies such as textbooks, problems, group work, or electronic feedback;

- investigating the difficulties of changing instructional paradigms, including the role of instructor beliefs and values, institutional constraints, student expectations, and student backgrounds; and
- investigating the role of basic thought processes in learning physics.

### Chemistry Education Research

In 1991, a groundbreaking article introduced what is now known as "Johnstone's Triangle" (Johnstone, 1991), which portrays the three central components of chemistry knowledge: the macroscopic, particulate, and symbolic (letters, numbers, and other symbols used to succinctly communicate chemistry knowledge) domains. These three domains have since provided a structure for chemistry education research. Indeed, questions about what students of chemistry know, or how teachers of chemistry ought to teach, mirror the quest of chemists to connect the macroscopic properties (color, smell, taste, solubility, etc.) of matter to the structure and particulate nature of matter.

Current areas of interest in chemistry education include

- students' conceptual understanding, especially of the particulate nature of matter (see Chapter 4);
- the use of technology to shape student reasoning;
- analysis of student argumentation patterns;
- the use of heuristics in student reasoning; and
- the development of assessment tools to measure thinking about chemistry (see Chapter 7).

### Engineering Education Research

Guided by the ABET accreditation criteria (ABET, 2009) and their implementation, the principal areas of inquiry for engineering education research include the following:

- the extent to which engineering education reflects engineering approaches by integrating and aligning content, assessment, and pedagogy for learning module, course, and program design (the equivalent of developing requirements or specifications, assigning relevant metrics, and preparing prototypes that meet the requirements) and by engaging in a cycle of improvement that closes the loop between research and practice;
- the extent to which engineering faculty adopt evidence-based practices;

- the extent to which faculty take a scholarly approach to teaching and learning or envision a developmental process for learning and inquiry;
- the extent of collaboration with higher education researchers, learning scientists, and other scholars of teaching and learning;
- the implicit and explicit values that departmental, college, and university cultures place on teaching and learning compared with traditional disciplinary research;
- the balance that Ph.D. programs strike between disciplinary research and the development of teaching and learning knowledge and skills;
- how engineers understand the nature of engineering work, especially early in their careers, but also across the career span; and
- strategies for helping students develop an understanding of what it means to be, and to become, an engineer.

As discussed in Chapter 2, these areas of inquiry and the ABET-defined areas of knowledge and skill development for engineering students have provided a framework for engineering education research since the late 1990s. One particular area of emphasis has been students' understanding of engineering concepts (Svinicki, 2011), with a concomitant focus on methods to promote greater conceptual understanding. Engineering education research also investigates methods for improving students' problem-solving and design skills.

Engineers pursue solutions to problems or improvements in the current state of the art, and engineering education researchers do the same. In aeronautical engineering courses, for example, prototypes such as sailplanes are used to demonstrate conceptual understanding, higher order thinking skills, and other dimensions of learning (Hansen, Long, and Dellert, 2002). However, these outcomes are not the focus of the research *per se*. Instead, engineering education research in this instance attends to how well the curriculum and instruction prepares students to understand the complexities of aeronautical engineering. The goal of preparing students for the future also highlights the importance of translating skills learned in the classroom to the workplace, which is another concern of engineering education research.

Some skills that are emphasized in ABET—teamwork, communication, and ethics/professionalism—are important in the engineering workplace, but have received relatively little attention from the engineering education research community. The awareness skills identified by ABET (appreciation for the impact of engineering on society locally and globally, commitment to lifelong learning, knowledge of contemporary issues) have received similarly little research attention.

## Biology Education Research

Since the mid-1990s, biology education research has followed the lead of physics education research by identifying students' conceptual understanding, building concept inventories, and assessing the effects of instructional interventions such as increased classroom engagement and group problem solving on students' learning (Dirks, 2011). Biology is a quantitative science, yet many students with math phobia enroll in biology, rather than other science courses, either to fulfill general education distributions or as a major. Thus, a current challenge for biology education researchers is to identify instructional approaches that can help overcome the math phobia of many biology students and introduce more quantitative skills into the introductory curriculum, as computational biology and other mathematical approaches become more central to the field of biology (National Research Council, 2003).

## Geoscience Education Research

Defining the scope of geoscience education research presents a challenge because there is no central "canon" of knowledge that is encompassed by the disciplines that study the earth (geology, oceanography, geophysics, geochemistry, atmospheric science, meteorology, climatology, planetary science, and physical geography). Geoscience content may be taught in a variety of courses, in different departments.

In the balance between implementing research findings to improve educational practice and accruing more such findings, geoscience education research has, to date, heavily emphasized the former. However, following other fields of DBER, geoscience education research built its first body of research around students' understanding of basic topics. These topics include the seasons, land forms, geological time, and natural hazards (Dahl, Anderson, and Libarkin, 2005; DeLaughter, Stein, and Bain, 1998; Kusnick, 2002; Libarkin, Kurdziel, and Anderson, 2007; Shepardon et al., 2007). Current areas of active inquiry include spatial thinking, temporal thinking, systems thinking, and field-based teaching and learning (Kastens, Agrawal, and Liben, 2009). In spatial thinking (Liben and Titus, 2012), geoscience education research finds common ground with geography education research (National Research Council, 2006), and in systems thinking (Stillings, 2012) with biology education research. Temporal thinking (Cervato and Frodeman, 2012; Dodick and Orion, 2006) and field-based learning (Maskall and Stokes, 2008; Mogk and Goodwin, 2012) appear at present to be distinctive to geoscience education research, with some parallel work in biology education research. Research on climate change education is an emerging interdisciplinary field (Gautier, Deutsch, and Rebich, 2006;

Marx et al., 2007; Mohan, Chen, and Anderson, 2009; Rebich and Gautier, 2005; Sterman and Sweeney, 2007; Weber, 2006), and an interesting example of the interplay between DBER and societal challenges.

### Astronomy Education Research

To date, astronomy education research has predominantly identified students' conceptual understanding. Another prominent focus of early research in astronomy education has been to address questions of overall teaching effectiveness (Bailey, 2011).

## METHODS

The methods DBER scholars use are as diverse as the research questions they investigate. Depending on the focus of the research, these methods range from qualitative interview studies or classroom observations of a few or dozens of students, to quasi-experimental comparisons of the learning of hundreds of students in similar courses across multiple institutions, to experimental manipulations in a research setting.

In some cases, the methods used by DBER scholars reflect the influence of the parent discipline. For example, astronomy is a quantitative science conducted by scholars with formal training in quantitative scientific methods, and the early history of astronomy education research was similarly dominated by quantitative research. Only recently has astronomy education begun to address questions similar to those pursued in the behavioral and social sciences, including questions that are best answered with qualitative methods (Bailey, Slater, and Slater, 2010). This trajectory of methodological approaches is similar to physics education research, and the trend to include a more robust combination of quantitative and qualitative studies is evidence that astronomy education research is maturing. Biology education research is another DBER field that is newly emerging from a quantitative discipline. As a result, the preponderance of biology education research is quantitative, and includes a relatively strong emphasis on quasi-experimental studies. In contrast, while experimental design is the norm in chemistry, chemistry education research has a long history of incorporating a wider range of qualitative and quantitative methods than are typically used in the parent discipline.

### Research Settings and Study Populations

Across the disciplines in this study, DBER scholars have studied similar types of courses. Despite the overall similarity of courses studied, however, not all institutions or student populations are equivalent in terms of class size, social background, and institutional priorities. These variations can have profound effects on outcomes and are important to consider when assessing the inferences that can be made from DBER findings.

## Research Settings

Large introductory courses are the primary setting for research in all DBER fields because these courses reach the most students. Research on student learning in these courses is often spurred by and related to the traditional overemphasis on memorization of factual information in a discipline, with an accompanying lack of student interest, shallow conceptual understanding, and poor retention (Sundberg, Dini, and Li, 1994).

Despite the prevalence of laboratory courses in the sciences and engineering and despite the importance of fieldwork in biology and the geosciences, very little DBER has been conducted in those settings. Moreover, relatively little research has been conducted in graduate or advanced-level undergraduate courses. Most of the latter comes from physics (e.g., Baily and Finkelstein, 2011; Pollock et al., 2011; Smith, Thompson, and Mountcastle, 2010) and chemistry (Bhattacharyya and Bodner, 2005; Orgill and Bodner, 2006; Sandi-Urena et al., 2011).

Some DBER has been conducted in the K-12 setting. Early research on learning and teaching chemistry, for example, investigated K-12 students because it was conducted by faculty who supervised preservice teacher training. Over time, chemistry education research came to include postsecondary students as faculty who taught introductory courses in chemistry departments began conducting research on those courses.

Conducting and interpreting research in introductory courses poses a number of challenges. A particular challenge in introductory biology courses is the breadth of the various divergent biology subfields, which further encourages broad, shallow introductory surveys of the discipline and hampers development of conceptual assessments that measure general biological knowledge across subfields of biology. In addition, the different subfields rely to some extent on different methodologies, for example the observational field work in ecology and the experimental laboratory research of molecular biology.

In contrast, astronomy education research has been motivated largely by a desire to improve teaching and learning in a single undergraduate course: the general education, introductory, nonmathematically oriented astronomy survey course known colloquially as ASTRO 101. The challenges of conducting research on ASTRO 101 and introductory geoscience courses are similar. In both disciplines, introductory courses typically include students who have little or no background in the subject and who usually are not considering careers in the discipline; undergraduates in

ASTRO 101 are most often future teachers or nonscience majors. Thus, faculty members are compelled to make these courses attractive, accessible and relevant to recruit and retain majors to the discipline, which means that the goals for these courses are often diffuse and broad. Moreover, ASTRO 101 is "terminal" in nature, rarely serving as a prerequisite for upper level courses. Because of these factors, introductory courses in the geosciences and astronomy can vary widely within and across institutions, posing a challenge for developing a coherent body of research on learning in these courses.

*Study Populations*

Given the focus of DBER on introductory courses, most studies include a mix of majors and nonmajors. Even in studies that investigate the conceptual understanding of individual students rather than the effectiveness of instruction as a whole, study participants typically are drawn from the enrollment in an introductory course. Majors and nonmajors in an introductory course can differ along many dimensions, including their motivations for taking the course, the extent to which they consider the course to be relevant to their studies and their futures, and their goals for learning and achievement. DBER studies do not always measure or explain these factors, which could play a role in learning. Further, as the following chapters show, very little DBER analyzes issues of teaching and learning as they relate to any different subpopulations of students. Although these limitations to the applicability of findings are not always explicitly acknowledged in DBER studies, they should be considered when drawing inferences from the research.

## THE ROLE OF LEARNING THEORIES AND PRINCIPLES

The extent to which DBER is grounded in broader theories and principles of learning and teaching varies widely. Many DBER studies either do not situate themselves in a broader theoretical frame, or do not explicitly define that frame. However, whether stated implicitly or explicitly, across the disciplines DBER is heavily influenced by constructivist ideas of learning, which propose that students generate understanding and meaning through experience (Ausubel, 2000; Dewey, 1916). Some DBER studies on collaborative learning are also influenced to varying degrees by sociocultural learning perspectives, which argue that students generate meaning and understanding by interacting in groups that share a common interest and learn together (Lave and Wenger, 1991), or through cognitive apprenticeships, where experts make tacit processes more explicit for novices

(Brown, Collins, and Duguid, 1989). The extent to which DBER studies use these perspectives to explain or extend their findings typically is limited.

The different fields of DBER approach the role of theory differently. Physics education research has strong ties to cognitive science research (Docktor and Mestre, 2011). Indeed, many cognitive science studies have investigated problem solving and the use of representations in physics, typically examining students' cognitive processing principles and internal mental processes (Bassok and Novick, 2012).

As with chemistry more broadly, the symbiosis of theory and measurement shape chemistry education research. The role of theory in experiment design is central to chemistry—data either support or refute theory—and theory plays a similarly important role in chemistry education research. Several resources have been published detailing how learning theory (Bretz and Nakhleh, 2001), methodologies (Orgill and Bodner, 2007), and experimental design in chemistry education research (Sanger, 2008; Towns, 2008) are grounded in the intersection of chemistry with several other disciplines.

In engineering, the Foundation Coalition, with funding from the National Science Foundation, undertook one of the few efforts to tie the ABET accreditation criteria to cognitive theories of learning. These efforts were designed to make the ABET criteria actionable and ground them in broader research. The coalition used Bloom's taxonomy of learning domains to develop a conceptual map linking ABET student learning criteria with learning objectives in the cognitive, affective, and psychomotor domains; assessments of those objectives; theories of cognition; and instructional approaches (see McGourty, Scoles, and Thorpe, 2002).

As discussed in Chapter 1, and as is evident from the synthesis in Chapters 4 through 7, DBER overlaps conceptually and theoretically with science education, educational psychology, cognitive science, and educational evaluation. More explicitly situating DBER in learning theories and principles from these fields would help to advance the conversations about teaching and learning in a given discipline, and in science and engineering more broadly. These principles and theories could explain some DBER findings, extend others, and form the foundations for deeper study.

## STRENGTHS AND LIMITATIONS

As with all research, DBER has strengths and limitations. DBER's greatest strength is its contribution of deep disciplinary knowledge to questions of teaching and learning. This knowledge has the potential to guide research that is focused on the most important concepts in a discipline, and offers a framework for interpreting findings about students' learning and understanding in a discipline. In these ways, even as an emerging field of inquiry, DBER has deepened the collective understanding of undergraduate learning in the sciences and engineering. When explicitly leveraged, the overlap of DBER with research from K-12 science education, educational psychology, and cognitive science can highlight findings that appear to be robust across different disciplines and learning contexts, and can help to identify differences that merit further exploration.

As described in Chapter 1, two of the long-term goals of DBER are to understand how people learn the concepts, practices, and ways of thinking of science and engineering and to help identify approaches to make science and engineering education broad and inclusive. Meeting these goals begins with an understanding of similarities and differences among different groups of students, yet very little DBER focuses on different subpopulations of students. At a time when the undergraduate population is becoming increasingly socially, economically, and ethnically diverse, a rich opportunity exists to enhance the understanding of the learning experiences of different groups. In a related vein, DBER could paint a more complete picture of undergraduate learning by taking into account differences among majors and nonmajors in introductory courses and structural differences among introductory courses, service courses for majors in other disciplines, and courses for majors.

At this point, DBER faces some challenges to the goal of independent reproducibility of research findings. Many DBER findings have been generated by the faculty members who are implementing the innovations and who developed the instruments to assess those innovations. The potential for investigator bias exists in these cases because these scholars naturally have a vested interest in the research results. One approach to counter this bias is to study other instructors who are implementing the innovation in question. However, it can be difficult to recruit others to teach specific course content in specific ways, independently of the research team.

Similar to other education research, the scale of most DBER studies poses a challenge to generalizing results, and to translating research findings into practice. A considerable proportion of DBER has been conducted at the scale of a single course, using instruments developed to assess learning in that course. As described elsewhere in this chapter, the variation in introductory courses across a discipline poses challenges to studying learning across those courses. Moreover, to the extent that the studies rely on instruments designed to measure student learning in the context of a single course, they might reflect standard examinations for that course. Such instruments generate little insight into broader issues of student learning, and limit the extent to which findings are applicable to other settings.

DBER has made some progress in addressing these challenges. For example, in the more established fields of DBER, such as physics and chemistry, scholars are developing instruments that can be widely used to generate deeper insights into students' understanding and learning experiences. And although multi-institutional studies are not the norm in DBER, they do exist. Part II of this report highlights these developments by describing the nature and quality of the existing evidence from discipline-based education research in physics, chemistry, engineering, biology, the geosciences, and astronomy, and synthesizing those literatures.

## ORGANIZATION OF THE SYNTHESIS

Across the next three chapters, we examine the literature on undergraduate students' conceptual understanding (Chapter 4), problem solving and use of representations (Chapter 5), and instructional strategies to improve science and engineering learning (Chapter 6). We devote a subsequent chapter (Chapter 7) to several emerging topics for DBER: science and engineering practices, applying knowledge in different settings (transfer), metacognition, and students' dispositions and motivations to study science and engineering (the affective domain).

Many of the topics in these chapters have been extensively studied in cognitive science, psychology, and science education. Our synthesis draws on relevant theoretical frameworks and findings from those disciplines to explain, extend, and contextualize DBER, while highlighting DBER's unique contribution of deep disciplinary knowledge to the understanding of these topics.

In reading Chapters 4 through 7, it is important to keep in mind that the nature of engineering and engineering education, combined with the strong influence of the ABET accreditation criteria on engineering education research, distinguish engineering education research from the other disciplines in this study. As a result, the body of engineering education research does not fit neatly into the categories around which we have organized the synthesis of the literature. As one example, because engineering education research emphasizes the integration and alignment of content (or curriculum), assessment, and pedagogy, it is difficult to identify studies in engineering that examine the efficacy of specific instructional strategies—the main focus of Chapter 6. We have parsed the engineering education research to fit the organization of this report, and Table 3-1 maps the ABET criteria onto the major sections of Chapters 4 through 7. Because the research base did not support a discussion of all ABET criteria, the report only discusses the criteria for which there are relevant, peer-reviewed studies.

**TABLE 3-1** Mapping ABET Student Learning Criteria onto Major Sections of the DBER Synthesis

| ABET Criteria (ABET, 2009, p. 3) | Applicable Sections of the Report |
|---|---|
| A: Ability to apply knowledge of mathematics, science, and engineering | Conceptual Understanding and Conceptual Change (Ch. 4) |
| B: Ability to design and conduct experiments, as well as to analyze and interpret data | The Role of Visualization and Representation in Promoting Conceptual Understanding and Problem Solving (Ch. 5) |
| | Metacognition (Ch. 7) |
| | Transfer (Ch. 7) |
| C: Ability to design a system, component, or process to meet desired needs | Problem Solving (Ch. 7) |
| D: Ability to function on multidisciplinary teams | Science and Engineering Practices (Ch. 7) |
| | Metacognition (Ch. 7) |
| E: Ability to identify, formulate, and solve engineering problems | |
| I: Recognition of the need for, and an ability to engage in lifelong learning | Dispositions and Motivation to Study Science and Engineering (Ch. 7) |
| F: Understanding of professional and ethical responsibility | Science and Engineering Practices (Ch. 7) |
| | Transfer (Ch. 7) |
| G: Ability to communicate effectively | |
| H: Understanding of the impact of engineering solutions in a global and societal context | |
| J: Knowledge of contemporary issues | |
| K: Ability to use the techniques, skills, and modern engineering tools necessary for engineering practice | |

# 4

# Identifying and Improving Students' Conceptual Understanding in Science and Engineering

One way to conceptualize undergraduate education is as a process of moving students along the path from novice toward expert understanding within a given discipline. To achieve this goal, it is important to begin by identifying what students know, how their ideas align with normative scientific and engineering explanations and practices (i.e., expert knowledge), and how to change those ideas that are not aligned.

Undergraduate science and engineering learning, like all learning, occurs against the backdrop of prior knowledge that students bring to the learning experience. Chi (2008) presents three levels of prior knowledge. In some situations, students may have no prior knowledge of the topic at hand. For example, at the start of a semester, students in an introductory Earth science class know the general concept of time yet have no knowledge about the significance of the geologic time periods. In such situations, learning can be viewed as adding new knowledge. In other situations, students may have correct but incomplete knowledge. For example, students in an introductory chemistry class may remember that the periodic table of the elements is arranged such that the elements in a particular column all have similar chemical properties, but that might be the extent of their knowledge about the information to be found in the periodic table. In these cases, learning can be conceived of as filling in the gaps. Finally, students may have incorrect knowledge that conflicts with the material to be learned, such as when students in an introductory biology class believe that lizards are more closely related to frogs than to mammals (Morabito, Catley, and Novick, 2010). In this case, learning involves conceptual change.

Research indicates that students at all levels, from preschool through college, enter instruction with various commonsense but incorrect interpretations of scientific and engineering concepts and skills (e.g., Chinn and Brewer, 1993), such as the well- known misconception[1] that the change in seasons is caused by changes in Earth's distance from the sun, rather than the tilt of Earth's axis (Schneps and Sadler, 1987). Some of these ideas are more firmly rooted than others, and thus are more resistant to change (Vosniadou, 2008a).

This chapter focuses on what is known about college students' conceptual understanding of science and engineering. To place discipline-based education research (DBER) in context, the chapter begins with a brief consideration of the broader knowledge base on students' conceptual understanding, including different theoretical perspectives. The chapter then summarizes DBER on conceptual understanding and on instructional practices to promote conceptual change and concludes with a summary of the key findings and directions for future research.

In this and subsequent chapters, the committee uses expert-novice differences and understandings as a framework for conceptualizing DBER findings. However, we recognize that expertise lies on a continuum, and we were guided by a relevant maxim from cognitive science that it takes 10 years for someone to acquire expertise in a domain (e.g., Ericsson, Krampe, and Tesch-Römer, 1993). Students are not expected to become experts within a single class, or even across the four years of their undergraduate education. They are, however, expected to progress along the path of increasing expertise. Thus, our frame of reference for this discussion is focused on helping students move toward the more expert end of the continuum.

## DIFFERENT PERSPECTIVES ON CONCEPTUAL UNDERSTANDING

Understanding what students know about science is the focus of considerable inquiry in cognitive science, educational psychology, and K-12 science education research (National Academy of Sciences, National Academy of Engineering, and Institute of Medicine, 2005; National Research Council, 1999, 2007). A key principle emerging from this research is that:

---

[1]In this report, we use the term "misconceptions" to mean understandings or explanations that differ from what is known to be scientifically correct. We recognize that other research refers to these explanations as "alternate conceptions," "prior understandings," or "preconceptions," and that the different terms can reflect different perspectives. When we use the term "misconceptions," we are following the convention of most DBER on this topic.

Humans are viewed as goal-directed agents who actively seek information. They come to formal education with a range of prior knowledge, skills, beliefs, and concepts that significantly influence what they notice about the environment and how they organize and interpret it. This, in turn, affects their abilities to remember, reason, solve problems, and acquire new knowledge. (National Research Council, 1999, p. 10)

Not all of students' ideas align with accepted science and engineering explanations, even if they are sensible and rooted in experience (National Academy of Sciences, National Academy of Engineering, and Institute of Medicine, 2005). Some research has focused on categorizing incorrect knowledge. In this regard, Chi (2008) argues that incorrect knowledge can be assigned to one of three levels, and that the approach to changing incorrect knowledge depends on the level of that knowledge:

1. Incorrect beliefs at the level of a single idea. An example is the false belief that all blood vessels have valves (Chi, 2008). In situations such as these, refutation might help students to change their beliefs.
2. Flawed mental models representing an interrelated set of concepts. For example, many students have a mental model of the human circulatory system as a single loop, rather than the correct model of a double loop (Chi, 2008; Pelaez et al., 2005). In these types of cases, multiple incorrect beliefs need to be corrected, ideally leading to the transformation of students' mental models. Although instruction is often successful in promoting such transformation, some students may instead assimilate correct concepts into their flawed mental model when those particular concepts do not directly contradict their model (also see Chinn and Brewer, 1993).
3. Assignment of core concepts to laterally or ontologically inappropriate categories. Examples of this type of incorrect knowledge include categorizing mushrooms as nonliving rather than living or believing that force is a substance-like entity that can be possessed, transferred, and dissipated, rather than a process. Such misconceptions have been found to be highly robust and resistant to change (Chi, 2005). In these cases, instruction needs to be focused at the categorical level, first teaching students the nature of the relevant categories so they can understand the concept in question as a member of that appropriate category.

Chi's (2008) tripartite taxonomy represents an eclectic approach to thinking about students' incorrect knowledge. Although she makes the forceful claim that most (perhaps all) robust misconceptions are due to lateral or ontological miscategorizations, other researchers are unconvinced that all instances of robust misconceptions can be classified as categorical mistakes. Indeed, a vibrant current area of research in cognitive science concerns the nature of students' initial, incorrect understandings of scientific concepts and phenomena (see Vosniadou, 2008b, for a thorough review of this literature).

One perspective is the "theory view," which suggests that students' concepts in a particular domain are coherent, systematic, and interrelated, essentially having the status of a naïve "theory." Although proponents of this view take different stances on the nature of such naïve theories, they share the view that students' knowledge is coherent (Vosniadou, Vamvakoussi, and Skopeliti, 2008). A contrasting perspective, the "pieces" view, proposes that students' naïve concepts are fragmented, piecemeal, and highly contextualized (diSessa, 2008). Although there may be some coherence across the numerous pieces of knowledge, this coherence does not rise to the level of even a naïve theory. Of course, these different perspectives on the nature of students' intuitive scientific knowledge may both be true, but for different areas of science. diSessa, Gillespie, and Esterly (2004) suggest that the extent to which to everyday experiences are connected to a particular set of scientific beliefs may be relevant, with the naïve theory view being more relevant when experiential knowledge is low and the pieces view being more plausible when it is high.

## OVERVIEW OF DISCIPLINE-BASED RESEARCH ON CONCEPTUAL UNDERSTANDING

Similar to scholars in other fields, DBER scholars have devoted considerable effort to identifying, documenting, and analyzing students' conceptual understanding (and misunderstandings). Indeed, investigations into the causes of students' reasoning difficulties and inaccurate beliefs about the physical world have dominated the physics education research literature since the 1970s (see Bailey and Slater, 2005, and Docktor and Mestre, 2011, for reviews, and see McDermott and Redish, 1999, for a list of approximately 115 studies related to misconceptions in physics). Likewise, with approximately 120 chemistry papers published on this topic between 2000 and 2010, students' conceptual understanding is one of the most active lines of inquiry in chemistry education research (see Barke, Hazari, and Yitbarek, 2009[2]). Considerably less research has been conducted on students' conceptual understanding in engineering (16 studies published between 2000 and 2010 as identified by Svinicki, 2011), biology (17 studies published between 2001 and 2010 as identified by

---

[2]For a web-based bibliography of students' conceptions, see http://www.ipn.uni-kiel.de/aktuell/stcse/stcse.html [accessed March 26, 2012].

Dirks, 2011; see Tanner and Allen, 2005 for a review[3]), the geosciences (79 studies published between 1982 and 2010 as identified by Cheek, 2010), and astronomy (Bailey and Slater, 2005). Although most of these studies focus on courses taken by majors and nonmajors in the first two years of college, a limited body of research also exists on upper division and graduate courses.

## Research Focus

For each field of DBER, initial research in this area often has involved cataloguing incorrect understandings and beliefs and identifying those that are more difficult to change than others. Across the disciplines, much of this research is predicated on the assumption that instructors need to know what their students already know, because prior knowledge can either interfere with or facilitate new learning (National Research Council, 1999). This research is sometimes coupled with instructional techniques that are designed to move students toward a more accurate understanding of the concepts at hand (see "Instructional Strategies to Promote Conceptual Change" in this chapter). When linked to the primary instructional goals of a discipline, these efforts represent an important first step in improving student learning. Most DBER on conceptual understanding in engineering, biology, the geosciences, and astronomy education research currently has this focus. As research on conceptual understanding within a discipline progresses, researchers seek linkages among the existing catalog of misconceptions. Identifying an underlying structure allows for the eventual development of more general instructional strategies that have the potential to address large classes of misconceptions rather than addressing them one at a time. Research on student understanding and knowledge construction in chemistry and physics have this focus. In physics two examples of this research are facets and P-Prims (diSessa, 1988; Minstrell, 1989), both of which are based on the perspective that student knowledge is characterized by fragmented pieces perspective. In chemistry, a different perspective— Ausubel, Novak, and Hanesian's (1978) construct of meaningful learning— has proved useful to identify fragments of information that students have memorized but not connected in a coherent, conceptual framework.

---

[3]Several web-based compilations of biology also exist. See for example, http://departments. weber.edu/sciencecenter/biologypercent20misconceptions.htm and http://teachscience4all. wordpress.com/2011/04/08/aaas-science-assessment-beta-items-for-assessing-misconceptions/ [accessed February 28, 2012].

## Methods

DBER scholars use a variety of assessment tools and research methods to measure students' conceptual understanding. These tools and methods include concept inventories (CIs); indepth interviews, concept maps, and concept sketches; surveys; and observations of students. In this section, we describe these methods, their strengths, and their limitations in terms of generating insights into students' understanding of concepts that are central to a discipline. Because these methods are commonly used to study a variety of topics in DBER —including those discussed in Chapters 5, 6, and 7—we elaborate on them here.

### Concept Inventories

CIs are used to assess students' preconceptions, to measure changes in response to a particular treatment, and to compare learning gains across individual courses dealing with a particular area of the discipline. Most commonly used in introductory courses, CIs are generally in a multiple-choice format, with incorrect responses (distractors) based on common misunderstandings or erroneous beliefs that have been identified by the literature (D'Avanzo, 2008; Libarkin, 2008). (See Box 4-1 for a description of one approach to developing a CI.) One exception is engineering, where the process of developing engineering CIs is generating much of the knowledge about student misconceptions in engineering, instead of the other way around (Reed-Rhoads and Imbrie, 2008).

With a long history in formative assessments (Treagust, 1988) and earlier lines of research, CIs initially gained traction in introductory physics with the development and widespread use of the Force Concept Inventory (Hestenes et al., 1992). CIs have since become increasingly common in other disciplines. In 2008, one researcher estimated that 23 CIs were in use across various science domains, with several others under development (Libarkin, 2008). The scope and quality of these CIs vary, as does the extent to which they have been validated.

Concept inventories (e.g., Mulford and Robinson, 2002) are not as widely used in chemistry as the Force Concept Inventory is in physics. In chemistry, misconceptions mostly have been identified by diagnostic assessments such as those developed by Treagust (1998) and conceptual exams developed by the American Chemical Society Examinations Institute.[4]

A particular strength of CIs is that their development often includes an identification of the most important concepts and learning goals for

---

[4]See http://chemexams.chem.iastate.edu for available examinations, study materials, and other resources related to these examinations [accessed March 25, 2012].

---

**BOX 4-1**
**Development of the Genetics Concept Assessment**

The following description summarizes the process of developing the genetics concept assessment, which is a concept inventory designed to measure learning gains from pre-test to post-test. The developers of this discipline-specific concept inventory followed the development process for concept inventories in other disciplines (Smith, Wood, and Knight, 2008, p. 423):

1. Review literature on common misconceptions in genetics.
2. Interview faculty who teach genetics, and develop a set of learning goals that most instructors would consider vital to the understanding of genetics.
3. Develop and administer a pilot assessment based on known and perceived misconceptions.
4. Reword jargon, replace distracters with student supplied incorrect answers [to pilot questions], and rewrite questions answered correctly by 70 percent of students on the pre-test.
5. Validate and revise the Genetics Concept Assessment through 33 student interviews and input from 10 genetics faculty experts at several institutions.
6. Give current version of the Genetics Concept Assessment to a total of 607 students, both majors and nonmajors, in genetics courses at three different institutions.
7. Evaluate the Genetics Concept Assessment by measuring item difficulty, item discrimination, and reliability.

---

a discipline or subdiscipline (Libarkin, 2008; Smith, Wood, and Knight, 2008; see Box 4-1). When they assess what experts in the field deem as central concepts, CIs provide a helpful structure for future research on conceptual understanding, and for the development of interventions to promote understanding that is aligned with normative scientific explanations. CIs are also useful because they can be used in large classes and across a range of students, allowing for greater generalizability of results; for longitudinal studies of the prevalence of certain misconceptions; and for disaggregating responses from multiple institutions along such dimensions as class size, type of institution, geographic setting, and demographic group (McConnell et al., 2006). However, as with all multiple-choice tests, CIs necessarily address a relatively coarse level of knowledge and provide no guarantee that a student who answers such a question understands the concept. Research has indeed shown that students may answer CI questions correctly even when they do not understand the concept (O'Brien, Lau, and Huganir, 1998). In addition, only erroneous ideas that are specifically targeted by the CI can be examined; CIs do not uncover new misunderstandings. Moreover, in the case of the Force Concept Inventory, scholars have debated about exactly what the CI is testing and the centrality of those concepts to the discipline (Hestenes and Halloun, 1995; Huffman and Heller, 1995).

### Interviews, Concept Maps, and Concept Sketches

DBER scholars commonly use indepth interviews to probe students' conceptions. Such interviews are typically conducted with one student at a time, and the typical sample size for most interview studies is fewer than 20. As in the social sciences, these interviews range from structured interviews with a fixed set of questions exploring a student's responses on a survey, CI, or other assessment of their understanding, to open-ended interviews that elicit students' thoughts and motivations, uncover common misconceptions, explore students' thinking processes, and examine their metacognition (see Chapter 6 for a discussion of metacognition or students' thinking about their learning processes).

DBER scholars also sometimes use concept mapping to assess conceptual understanding. Developed by Novak in the 1970s (Novak and Gowin, 1984), concept maps are designed to provide a nonlinear, two-dimensional impression of how students relate (and interrelate) a list of concepts. Typically the concepts in question are linked by words or phrases to indicate how they are related (see Figure 4-1 for an example). Concept maps and concept sketches have been used to assess conceptual understanding in chemistry (Lopez et al., 2011); for various purposes in engineering (Besterfield-Sacre et al., 2004; Heywood, 2006); and in the geosciences to measure conceptual change that has occurred after instruction and to reveal students' understanding of processes, concepts, interrelationships, and key features (Englebrecht et al., 2005; Johnson and Reynolds, 2005; Rebich and Gautier, 2005). Like all methodologies, concept maps have limitations. When used as assessment tools, they can be difficult to score and difficult to compare to a "correct" concept map because there is never just one correct concept map. Also, no inferences can be drawn from any ideas students omit from the map. And although the collective body of research using these tools generates insights into a wide variety of concepts, not all of these concepts are central to expert understanding of the discipline.

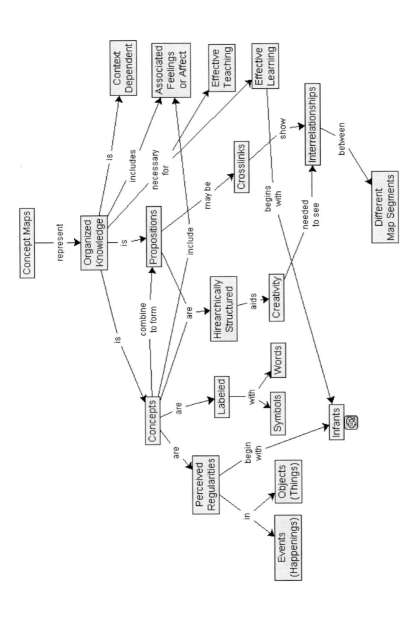

**FIGURE 4-1** Example of a concept map.
SOURCE: English Wikipedia user Vicwood40 (2005). Available: http://en.wikipedia.org/wiki/File:Conceptmap.gif [February 21, 2012; printed as posted].

*Surveys*

In-class surveys or surveys across multiple classes or institutions are another mechanism for understanding students' ideas and conceptual understanding. These surveys typically contain forced-choice and/or open-ended questions about specific aspects of the course content or learning experiences. Surveys mix the relative ease of implementation and analysis of concept inventories with the more open-ended nature of interviews. However, self-reported data can be unreliable—even with carefully designed surveys—so results must be interpreted with caution (Marsden and Wright, 2010). Moreover, the degree to which instructors use surveys that they develop themselves limits the degree to which results can be generalized.

Given their relative strengths and limitations, the research methods described above should be used in concert to bring out the full timbre of students' understanding, then combined with instruction to promote more expert-like understanding. An example from biology education research illustrates the limitations of relying solely on a single mode of assessment. In a study of misconceptions about blood circulation held by undergraduate biology students who were prospective elementary teachers, Pelaez et al. (2005) integrated methods of identifying common misconceptions with learning activities designed to deepen understanding. These assessment methods included pretest drawings, peer reviewed essays, debates that required students to integrate knowledge, written exams, and oral final exams that consisted of probing interviews. The authors concluded that "Multiple data sources were necessary to expose many errors about circulatory structures and functions. Drawings combined with individual interviews provided the richest source of information about student thinking. Relying solely on the essay exam would not have uncovered the magnitude of the problem" (Pelaez et al., 2005, p. 178).

## UNDERGRADUATE STUDENTS' UNDERSTANDING OF SCIENCE AND ENGINEERING CONCEPTS

For every discipline in this study, DBER has revealed that undergraduate students have misunderstandings and incorrect beliefs related to a wide range of concepts (Bailey and Slater, 2005; Barke, Hazari, and Yitbarek, 2009; Cheek, 2010; Dirks, 2011; Docktor and Mestre, 2011; McDermott and Redish, 1999; Svinicki, 2011; Tanner and Allen, 2005). Across the disciplines, students have difficulties understanding interactions or phenomena that involve very large or very small spatial scales (e.g., Earth system processes, the particulate nature of matter, quantum mechanics) and take very long periods of time (e.g., natural selection, Earth history). Considering that students often use their own experiences to generate scientific explanations, it stands to reason that they have difficulties with concepts for which they lack a frame of reference (National Academy of Sciences, National Academy of Engineering, and Institute of Medicine, 2005; National Research Council, 1999, 2007).

Although DBER on upper division and graduate courses is currently relatively limited, the available research suggests that many incorrect understandings and beliefs are highly resistant to change (Orgill and Sutherland, 2008; Rushton et al., 2008). For example, a body of research on thermodynamics shows that some incorrect beliefs students hold in high school, chiefly that chemical bonds release energy when they break, still remain after students complete several undergraduate courses in chemistry (Canpolat, Pinarbasi, and Sözbilir, 2006; Sözbilir, 2002, 2004; Sözbilir and Bennett, 2006). Organic chemistry students also have difficulties understanding hydrogen bonding—a topic that is fundamental to understanding the chemical properties of organic molecules, and that is first encountered in introductory chemistry (Henderleiter et al., 2001). In general, students are able to define hydrogen bonding, but they have trouble using hydrogen bonding to predict properties of molecules. Even more striking, some graduate students in chemistry doctoral programs still harbor confusion about phase changes—a topic first taught in elementary school (Bodner, 1991). A common misconception about phase changes is that bubbles in boiling water are made of air rather than of water vapor (Nakhleh, 1992).

As described in Chapter 1, a defining characteristic of DBER is deep disciplinary knowledge of the topic under consideration. For measuring students' conceptual understanding, this knowledge is vital to identifying concepts that are central to a given discipline, and identifying the expert-like understandings that are the goals of instruction on those concepts. Some, but not all, DBER studies on conceptual understanding involve concepts that are central to the discipline. As discussed under "Concept Inventories" several existing concept inventories have included an explicit identification of the concepts that are vital for students to gain a more expert-like understanding of the discipline. For example, engineering CIs are being developed in response to standards developed by the accrediting agency ABET (see Chapter 2); these CIs primarily address ABET criteria A (an ability to apply knowledge of mathematics, science, and engineering) (Reed-Rhodes and Imbrie, 2008).

Beyond CIs, some notable examples of research that is central to the discipline include the body of research on temporal and spatial scales in the geosciences (Catley and Novick, 2009; Cheek, 2010; Hidalgo, Fernando, and Otero, 2004; Teed and Slattery, 2011; Trend, 2000); seasons and moon phases in the Sun-Earth-Moon system in astronomy (Schneps and Sadler, 1987; Zeilik et al., 1997, 1999, 2000, 2002); and research on the three domains of Johnstone's triangle in chemistry, discussed next.

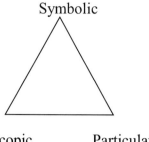

Symbolic

Macroscopic          Particulate

**FIGURE 4-2** Johnstone's triangle representing the three central domains of chemistry.

Johnstone's triangle (1982) portrays three central components of chemistry knowledge: the macroscopic, particulate, and symbolic (letters, numbers, and other symbols used to succinctly communicate chemistry knowledge) domains (see Figure 4-2). Because chemists expect students to develop fluency with all three of these domains and understand their connections to one another (Johnstone, 1982, 1991), assessing students' understanding of each domain is an important line of chemistry education research. That research has shown that students have trouble understanding all three domains, and that difficulties understanding the particulate nature of matter represent one of the most important barriers for students (see, for example, Gabel, Samuel, and Hunn, 1987; Yezierski and Birk, 2006).

Understanding the structure of matter at the particulate level is critical to understanding the behavior and interactions of molecules. During their general chemistry courses, students learn to construct Lewis structures, which show the arrangement of atoms, bonds, and electrons in a molecule. Faculty then use these structures to explain molecular shape and subsequently emphasize how shape and electron distribution influence physical properties such as polarity, solubility, and miscibility. However, as discussed in Chapter 5, several studies have shown that students have difficulty constructing Lewis structures, and their representations of molecules do not necessarily improve over time (Cooper et al., 2010; Nicoll, 2003).

## INSTRUCTIONAL STRATEGIES TO
## PROMOTE CONCEPTUAL CHANGE

When students harbor known misunderstandings and incorrect ideas or beliefs about concepts that are fundamental to a discipline, moving toward more expert-like understanding can involve conceptual change, or helping to align their beliefs with accepted science and engineering ideas. Teaching for conceptual change requires that instructors understand and explicitly address these everyday conceptions and help students to refine or replace them (National Academy of Sciences, National Academy of Engineering, and Institute of Medicine, 2005). This process goes beyond identifying misconceptions, to being aware of their origins (e.g., common wisdom or "folk science," the result of instruction, or cultural strictures such as religious faith), understanding their roots in deeper cognitive mechanisms, and considering their impact on student learning across disciplines.

Promoting conceptual change is challenging because it is a slow process and some ideas are more deeply rooted than others (Vosniadou, 2008a). In addition, incorrect ideas, beliefs, and understandings arise in many different ways, and their origins have implications for instruction. Some ideas arise because they align with personal experience. For example, the belief that denser objects fall more quickly than lighter objects in a vacuum is consistent with the observation that rocks fall more quickly than leaves (Docktor and Mestre, 2011). Some incorrect ideas are induced by instruction (Wandersee, Mintzes, and Novak, 1994), perhaps because students have no previous experience with the phenomenon in question—representing the atomic scale, for example (Taber, 2001)—because students overgeneralize from instructional analogies (Jee et al., 2010), or because there are inaccuracies in teaching materials (Hubisz, 2001). Indeed, although inaccuracies in K-12 teaching materials have been well-documented (King, 2010), similar studies have not been conducted for college-level teaching materials. Some scholars also claim that incorrect ideas about controversial topics such as the ozone hole, acid rain, and climate change have been deliberately fostered (Oreskes, Conway, and Shindell, 2008).

Effective conceptual change depends on students' understanding that their beliefs are hypotheses or models rather than facts about the world, that other people may have other beliefs/hypotheses/models, and that these hypotheses or models need to be evaluated in light of relevant empirical evidence (National Academy of Sciences, National Academy of Engineering, and Institute of Medicine, 2005). Thus, considerations of students' understanding of the nature of science and engineering and of the process of learning also are key to promoting conceptual change.

Clement (2008) discusses a range of types of conceptual change such as modifying an existing model by adding, removing, or changing elements; creating a new model (that has not grown out of an existing model); or replacing a concept with an ontologically different concept. He argues that all of these types of conceptual change are applicable at times, so a variety of different teaching strategies will be needed, possibly even within a single class.

Several approaches have been used to promote conceptual change in physics (see Docktor and Mestre, 2011, for a review). The University of Washington physics education research group identifies misconceptions and then engages in a cyclic process of designing an intervention (based on previous research, instructor experiences, or expert intuition), testing and evaluating the intervention, and then refining the intervention until evidence is obtained that the intervention works (McDermott and Shaffer, 1992).

One instructional strategy is to use "bridging analogies" that provide a series of links between a correct understanding that students already possess and the situation about which they harbor an erroneous under-standing (Brown and Clement, 1989; Camp, Clement, and Brown, 1994; Clement, 1993; Clement, Brown, and Zietsman, 1989). Research in chem-istry (Zimrot and Ashkenazi, 2007) and physics (Sokoloff and Thornton, 1997) also has shown that interactive lecture demonstrations can promote conceptual change. For example, Sokoloff and Thornton (1997) found that after students experienced an interactive lecture demonstration related to Newton's Third Law, they retained appropriate understanding of the law months later. Effective problem-solving instruction more broadly also has been shown to promote conceptual change (Cummings et al., 1999; see Chapter 6). On the other hand, as an example of the difficulty of promot-ing lasting conceptual change, college students have been shown to perform well on exams on the laws of motion, but continue using their incorrect, experience-based ideas to act in the world (diSessa, 1982).

A limited amount of engineering education research has focused on assessing conceptual change (see Turns et al., 2005). One strength of this research lies in the cooperation between engineering education researchers and educational psychologists to incorporate effective measurement and assessment techniques in the study of engineering education outcomes. Although the committee has characterized the strength of findings from this research as limited because few studies exist and most were conducted on a small scale, the findings point to increased conceptual understanding, in general, over time among students in engineering programs. However, in addition to being limited, the evidence is mixed and does not indicate how long the conceptual change lasts. Some studies demonstrate positive changes in conceptual knowledge over time (e.g., Muryanto, 2006; Segalas, Ferer-Balas, and Mulder, 2008) while others show limited change over the life of the academic program (e.g., Case and Fraser, 1999; Montfort, Brown, and Pollock, 2009). One possible explanation for this discrepancy is the degree to which programs develop concepts over time. Engineering design, like all complex subjects, requires repeated exposure rather than a single intense immersion or two (e.g., freshman engineering, capstone design) (Cabrera, Colbeck, and Terenzini, 2001). Indeed, cognitive science research, and science education research have shown that students are more likely to change their conceptions when they interact more with the content and the learning process (National Research Council, 1999, 2007; Stevens et al., 2008; Taraban et al., 2007; see Box 4-2 for an example).

---

**BOX 4-2**
**Changing Students' Conceptions About Heat Transfer**

Students often believe that heat is a substance rather than energy transfer. This misperception presents challenges throughout the chemical engineering curriculum. These misconceptions are particularly robust because they represent the mis-categorization of ontological differences. In one study, 23 introductory chemical engineering students engaged in three laboratory activities to address very specific heat transfer misconceptions. In the first activity, they worked with boiling liquid nitrogen in an experiment related to the misconception that temperature is a good indicator of total energy as opposed to average kinetic energy. The second laboratory activity compared heat transfer in chipped versus block ice to address the misconception that more heat is transferred if a reaction is faster. The third laboratory activity, which involved cooling heated blocks with ice, was designed to help students distinguish between rate and amount of heat transfer. The researchers compared students' scores on the heat transfer concept inventory before and after the laboratory activities and reported statistically significant gains on many of the items that pertained to the activities. These results suggest that conceptual change can occur, at least in the short term.

SOURCE: As discussed in Prince, Vigeant, and Nottis (2009).

---

Geoscience education research also has identified some strategies to successfully align students' beliefs with accepted scientific explanations; the committee characterized the strength of findings from this research as limited because few studies exist and they were typically conducted within individual courses over brief time frames. McNeal, Miller, and Herbert (2008) used inquiry-based learning and multiple representations to effect conceptual change regarding increased plant biomass caused by increased nutrients in coastal waters (coastal eutrophication). Other research has reported large increases in knowledge and a decrease in misconceptions immediately after a three-week mock summit on climate change that used "role-playing, argumentation, and discussion to heighten epistemological awareness and motivation and thereby facilitate conceptual change" (Rebich and Gautier, 2005, p. 355). As an example of the difficulty of promoting lasting conceptual change, other researchers examined students' concept maps over a two-semester sequence of introductory geology lectures and found an increase in the number of geological concepts identified but "a disproportionately small increase in integration of those concepts into frameworks of understanding" (Englebrecht et al., 2005, p. 263).

## SUMMARY OF KEY FINDINGS ON CONCEPTUAL UNDERSTANDING

- *In all disciplines, students have incorrect ideas, beliefs, and explanations about fundamental concepts. These ideas pose challenges to learning science and engineering because they are often sensible, if incorrect, and many are highly resistant to change.*
- *Many robust misunderstandings and incorrect beliefs have been identified, but not all are equally important. The most useful research focuses on ideas, beliefs, and understandings that involve central concepts in the discipline and that are widely held.*
- *In general, students have difficulty understanding phenomena and interactions that are not directly observable, including those that involve very large or very small spatial and temporal scales.*
- *A variety of tools and approaches have been used to measure students' conceptual understanding, ranging from highly focused interviews to broader measures such as concept inventories. Although each tool has its own strengths and limitations, it is vital for them to address the key concepts and practices of a discipline.*
- *A variety of teaching strategies is needed to help students refine or replace incorrect ideas and beliefs, possibly even in a single unit of instruction. Physics education research has identified several strategies for successfully promoting conceptual change, including interactive lecture demonstrations, interventions that target specific misconceptions, and "bridging analogies" that link students' correct understandings and the situation about which they harbor a misconception.*

## DIRECTIONS FOR FUTURE RESEARCH ON CONCEPTUAL UNDERSTANDING AND CONCEPTUAL CHANGE

Across the disciplines considered in this report, a substantial body of literature exists about students' conceptual understanding. Nonetheless, many gaps remain. All disciplines would benefit from well-validated, overarching schemas that describe the kinds of phenomena about which humans are prone to develop misunderstandings; Talanquer's (2002, 2006) research on interpreting students' ideas in chemistry may be a step in this direction. This research should focus on the key concepts and practices that are central to learning in a given discipline. A thorough understanding also is needed of whether and how the types and persistence of misconceptions differ for students from different groups, including gender, race/ethnicity, academic ability, urban vs. rural, and majors versus nonmajors.

Perhaps even more importantly, with the exception of physics, very little research at the undergraduate level provides evidence of conceptual change over time as a result of instruction or other learning experiences. Many researchers provide suggestions for instruction, but fewer provide evidence about the efficacy of these suggestions (e.g., Tanner and Allen, 2005). Although physics education research has identified several pedagogical techniques that reliably move students toward scientifically normative conceptions, additional research is needed to understand whether these strategies work in other disciplines. Studies on these strategies in physics have been informed by cognitive science, and future DBER in other disciplines should be as well. For example, recent cognitive science research (Vosniadou, Vamvakoussi, and Skopeliti, 2008) has helped to replace Posner's (1982) classical approach to conceptual change with more nuanced, and competing, perspectives. These newer perspectives emphasize the continuity of knowledge that can be expected over the course of learning and thus the possibility of identifying elements of novices' prior knowledge that can contribute to the construction of more expert knowledge structures.

Research on effective methods for promoting conceptual change is particularly important because evidence exists for the persistence of incorrect ideas, beliefs, and explanations even after years of studying science and engineering. However, the length of many studies on this topic is often just one semester. Longitudinal studies of curriculum and pedagogy experiments designed to help students move toward normative scientific and engineering explanations are needed to more fully understand when and why incorrect ideas persist or reemerge. When applicable, these studies also should take into account the relationship between students' personal belief systems and conceptual change. Research suggests that although students may be able to effectively apply knowledge that is inconsistent with their personal beliefs (such as answering questions about evolution correctly while not accepting evolution as valid) (Blackwell, Powell, and Dukes, 2003; Champagne, Gunstone, and Klopfer, 1985; Sinatra et al., 2003), their awareness of their beliefs and willingness to question those beliefs affect their receptivity to conceptual change (Pintrich, 1999).

Another important question for further DBER is how specific incorrect ideas originate, as a way of identifying effective means of moving students toward more normative understanding. Future DBER in this area might be informed by the growing body of research in cognitive science that focuses on the nature of students' initial, incorrect conceptions of scientific concepts and phenomena (see Vosniadou, 2008b, for a thorough review of this literature). Progress on this set of questions would benefit from collaboration between DBER scholars and K-12 science education researchers, especially those working on learning progressions (Mohan, Chen, and Anderson, 2009; Plummer and Krajcik, 2010; Schwarz et al., 2009). Research on the conceptual understanding of pre- and in-service K-12 teachers (Dahl, Anderson, and Libarkin, 2005; Kusnick, 2002) also may be a fruitful area of common ground between the two research communities.

Finally, the committee identified some discipline-specific needs for future research. More information is needed in engineering, biology, and the geosciences to design assessments that can diagnose students' difficulties and to design instruction to move them toward more accurate understanding. Chemistry education could benefit from additional measures that specifically target students' conceptual understanding because only a few such measures exist—to date, chemistry education researchers have used a variety of other tools to uncover and document incorrect ideas and beliefs. In astronomy, the next generation of assessment instruments is emerging, including assessments for general astronomy (Slater, Slater, and Bailey, 2011) and for targeted topic areas such as stars and stellar evolution (Bailey et al., 2011), light and spectra (Bardar, Prather, and Slater, 2006), planetary science (Hornstein et al., 2011), and the influence of gravity (Lindell and Sommer, 2003). The goal of this next generation of instruments is to reveal the underlying cognitive processes undergraduates use when thinking about these topics in astronomy. Research also is needed on students' cognition and spatial thinking skills to more fully explore undergraduates' conceptual understanding in astronomy. In physics, research on the underlying cognitive processes learners use when engaging in physics and how these might change as learners progress from novice to more expert perspectives has already begun. Although this type of research is more difficult and time-consuming than cataloging misconceptions, and might not be immediately applicable to instruction, it has the potential to generate broader insights about student understanding that could be relevant to many disciplines.

# 5

# Problem Solving, Spatial Thinking, and the Use of Representations in Science and Engineering

Chapter 4 explored students' conceptual understanding in science and engineering, with the goal of helping students advance toward a more expert-like understanding. This chapter addresses how students use those understandings to solve problems, and how scientific representations, such as pictures, diagrams, graphs, maps, models, and simulations facilitate or impede students' problem solving and understanding of science and engineering. Although we recognize that there are other important dimensions to promoting a deep understanding of science and engineering—including strong mathematical knowledge—we start with these topics because they are vital to acquiring greater expertise in the disciplines, and discipline-based education research (DBER) on them is relatively extensive and robust.

The discussion of each topic in this chapter begins with an introduction of that topic and its importance to undergraduate science and engineering education. Following these introductions, we provide an overview that summarizes the focus of DBER on the topic, the theoretical frames in which DBER is grounded, and the typical methods used. We then discuss the research from each discipline and summarize key findings across disciplines. The discussion of each topic concludes with an identification of directions for future research.

## PROBLEM SOLVING

Problem solving may be the quintessential expression of human thinking. It is required whenever there is a goal to reach and attainment of that goal is not possible either by direct action or by retrieving a sequence of previously learned steps from memory (Bassok and Novick, 2012; Martinez, 2010). That is, during problem solving the path to the intended goal is uncertain. This characterization describes much of what people do on a daily basis, from (a) mundane activities like deciding what to cook for dinner given the ingredients at hand or how to get from work to home given certain street closures, to (b) student activities such as interpreting laboratory results, figuring out how to organize a term paper on evidence for speciation, or designing a roller coaster for an engineering class, to (c) professional work such as curing illnesses or determining the best way to structure a class so that students will understand a key concept. Clearly, problem solving is central to science and engineering as well to everyday life.

Researchers in numerous disciplines have drawn a distinction between well-defined and ill-defined problems (Hsu et al., 2004; Reitman, 1965). Most of the problems students encounter in their science and engineering classes are well-defined, such as a mechanics word problem. In these problems, the initial conditions, the goal, the means for generating and evaluating the solution, and the constraints on the solution are all clearly specified for students. For other types of problems, however, such as a more open-ended laboratory or an authentic design problem in engineering, students have to define one or more of the problem components on their own (Fay et al., 2007; Whitson, Bretz, and Towns, 2008). In a laboratory, the means of generating the solution may be ill-defined. For an engineering problem, the goal may be ill-defined; as a result, it may not be clear how to determine whether the goal has been accomplished. For example, what constitutes a better coffee cup, and how does one decide that a new cup design represents a big enough improvement over the status quo to declare the design finished?

Society's most important problems are usually ill-defined in some way. Consider two examples: (1) How can the rapid regrowth of human skin be promoted so that life-threatening infections in burn patients are prevented? (2) How can affordable, alternative energy to power cars be generated, thereby limiting reliance on fossil fuels? These are the kinds of problems students will have to solve after they graduate. Students who have scant experience with ill-defined problems during their undergraduate education may be poorly prepared to grapple with the most significant problems in their fields.

This discussion of problem solving is structured around important findings from DBER that are consistent with prominent themes from the cognitive science literature, namely problem representation and the nature of the solution process. In the cases for which the findings apply to only a small number of problem domains or disciplines, their broader applicability to problem solving within the disciplines of interest here is an open question. For example, as the following discussion will show, research has shown that experts adopt a working forward strategy in certain situations (e.g., solving introductory mechanics problems from physics). That strategy may reflect the nature of the particular problem-solving tasks that have been investigated; other problems, from the same or other disciplines, may require different strategies or approaches for successful and/or efficient solution.

Where findings have been replicated in numerous disciplines within and outside the sciences, it is probably safe to presume that those findings generalize to new problem domains or disciplines yet to be investigated. A prime candidate for such a finding is the differential reliance of experts and novices on structural versus superficial features of problems, respectively.

A potential complication of generalizing from cognitive science research relates to the research setting and nature of the problems studied. DBER is typically conducted in classroom settings with discipline-specific problems, whereas much of the cognitive science research—especially early in that field's history—has been conducted in laboratory settings with puzzle problems or brain teasers. However, cognitive science research on problem solving in more ecologically valid settings and in domains such as physics and mathematics has often yielded comparable results to studies of puzzle problems (Bassok and Novick, 2012). What changes from one problem to another in these situations is the specific knowledge students need to bring to bear on their solution attempts, rather than the underlying cognitive processes. This general pattern of consistent results across disparate types of problems lends support to the committee's view that findings from cognitive science research on problem solving may be applicable in undergraduate science and engineering domains in which they have not yet been investigated. After all, humans have a single cognitive system, with specific operating parameters and constraints, that underlies their learning and problem solving regardless of the problem or discipline under investigation (Simon, 1978). At the same time, domain knowledge, which the general processes take as input, is important as well. In the inevitable cases where different patterns of results are found across problems or disciplines, these patterns will point to specific areas of science learning where disciplinary knowledge and perspectives are especially critical.

## Overview of Discipline-Based Education
## Research on Problem Solving

Problem solving is a significant focus of DBER in physics (see Docktor and Mestre, 2011, for a review), chemistry (for reviews, see Bodner and Herron, 2002; Gabel and Bunce, 1994), and engineering (see Svinicki, 2011, for a review), and an emerging area of study in biology and the geosciences. Because problem solving is not taught frequently enough in astronomy, the committee did not find peer-reviewed astronomy education research on problem solving. As a result, this discussion does not include astronomy.

A significant body of research on problem solving also exists in cognitive science, and that research overlaps considerably with DBER. Cognitive science research corroborates some DBER findings, can help to explain or extend others, serves as the theoretical basis for some studies, and provides potential building blocks for future DBER on problem solving (Bassok and Novick, 2012). Because of these linkages, this section interweaves discussions of DBER and cognitive science.

*Research Focus*

DBER studies on problem solving range from investigations of general problem solving strategies, to behavioral differences between novices and experts, to measurements of the effectiveness of instructional strategies that teach problem solving. Most of these studies investigate how students solve quantitative, well-defined problems. Accordingly, unless otherwise noted, the bulk of the discussion in this chapter refers to well-defined problems.

The rich research base on problem solving in physics builds on many studies in cognitive science dating back more than 50 years. Many of these studies are based on the information-processing approach to understanding thinking, which comes from cognitive psychology (e.g., Simon, 1978). Key ideas from this framework include a step-by-step approach to problem solving, the importance of both internal knowledge representations and processes for understanding human thinking, the role of prior knowledge (which supports analogical transfer of knowledge gained from previously solved problems to solve new problems), and a limited capacity processing system.

In chemistry, the study of problem solving is muddied by disagreements over what constitutes a problem (Bodner, 2004). These debates notwithstanding, a large group of studies has examined problem solving strategies in a specific content area of chemistry, such as stoichiometry or equilibrium. Studies on these topics have used several models of problem solving as a framework for inquiry, including the Pólya model (1945)—originally developed in the context of mathematics—of understanding the problem, devising and carrying out a plan, and checking work; Wheatley's (1984) model of the steps successful problems solvers take to solve novel problems in mathematics; and the expert novice paradigm described in this chapter. More recently, chemistry education research studies have drawn on knowledge space theory, which describes possible states of knowledge (Taagapera and Noori, 2000) and the ACT-R theory for understanding human cognition (Taatgen and Anderson, 2008).

Most research on student learning outcomes in engineering focuses on problem solving and engineering design (ABET accrediting criteria C and D; see Chapter 3). Early studies on this topic drew on information processing theory (Simon, 1978). More recent studies are grounded in constructivist (Piaget, 1978) or, less commonly, socioconstructivist (Lave and Wenger, 1991; Resnick, 1991) theories of learning.

Research on problem solving in biology and the geosciences is sparse. The six biology studies that the committee reviewed examined individual differences in problem-solving strategies and did not explicitly situate themselves in broader learning theory. In the geosciences, one emerging line of research draws on the cognitive science field of naturalistic decision making (Klein et al., 1993; Marshall, 1995) to investigate student problem solving in the field setting using global positioning satellites.

## Methods

DBER scholars use a wide variety of qualitative and quantitative methods to study problem solving. Some data are gathered using think-aloud interviews, in which students are asked to solve problems and verbalize their thoughts while being video and/or audio taped.[1] Many studies comparing expert and novice problem solvers have used categorization tasks, which require participants to group problems based on the similarity of the solution method. Other methods of tracking student problem-solving strategies include computer systems that use knowledge space theory (Taagepera and Noori, 2000), and artificial neural networks and Hidden Markov Models (Cooper et al., 2008).

Some studies use student-generated summaries of their problem-solving approaches, course exam scores, and final grades to measure proficiency with problem solving, rather than examining problem solving processes. In physics, students' written solutions to problems that have been designed by the researcher(s) and/or adapted from existing problem sources such as textbook end-of-chapter problems are another common data source. These data are typically analyzed by scoring students' solutions relative to rubrics that characterize expert problem solving.

Study populations range from high school students to community college students to graduate students, with the preponderance of studies focusing on students enrolled in introductory college courses. Sample sizes in these studies range from fewer than 20 to several hundred students. The research typically is conducted in classroom settings—although in physics, this research also involves students both in research settings.

---

[1]Ericsson and Simon (1980, 1993) have provided important theoretical and practical guidance for collecting and interpreting think-aloud protocols.

Many DBER studies of problem solving have compared undergraduate students (novices) to more expert problem solvers such as graduate students, faculty, or professionals outside academia (e.g., Larkin et al., 1980; Petcovic, Libarkin, and Baker, 2009). Similarly, a major focus of cognitive science research on problem solving has been to compare the performance of novices (usually, although not always, college students) with that of experts. The definition of *expert* varies across studies, ranging from graduate students in an academic discipline such as physics, to grandmaster chess players or practicing physicians with 20 years of experience in their field. In cognitive science research, the typical study has used an extreme groups design, comparing a group of novices to a single group of (relative) experts. Fewer studies have compared problem solving across multiple levels of expertise. Regardless of the number of groups included, these studies provide important information for discipline-based education researchers because they give insight into the nature of the transition that needs to occur and the goal toward which students should strive (Lajoie, 2003).

As discussed in Chapter 4, the committee recognizes that students are not expected to become experts within a single class, or even across the four years of their undergraduate education. They are, however, expected to progress along the path of increasing expertise. Thus, our frame of reference for this discussion is focused on helping students move toward the more expert end of the continuum.

## The Nature of the Solution Process

A problem representation is an internal (i.e., existing in memory) or external (e.g., drawn on paper) model of the problem that is constructed by the solver to summarize his or her understanding of the problem. Ideally, this model includes information about the objects or elements in the problem, their interrelations, the goal, the types of operations that can be performed on the elements (e.g., algebraic operations for certain types of problems), and any constraints on the solution process. A student's representation of the problem at hand is critical because the representation constructed affects the types of operations that can be applied (i.e., the steps that can be taken) to solve the problem (see "The Role of Visualization and Representation in Conceptual Understanding and Problem Solving" in this chapter).

According to a review of cognitive science research (Bassok and Novick, 2012), for some problems, getting the right representation is the key to solving the problem, or at least to solving it in a straightforward manner. Although problem representation is especially important for ill-defined problems, it can also be critical for solving well-defined problems. For other problems, determining the best representation is a relatively straightforward process, and the primary work is to discover a (or the best) path connecting the situation as presented to the goal state. Representation and step-by-step solution are interactive processes, however, and both are important in most cases of problem solving. As noted, the solver's representation of the problem guides the process of generating a possible solution. The step-by-step solution process, in turn, may change the solver's representation of the problem, leading to corresponding changes in the solution method attempted. This iterative process of representation and step-by-step solution continues until the problem is solved or the solver abandons the goal.

One difference between relative experts and novices concerns how they allocate their problem solving time between creating a representation and working to find a solution. In some disciplines, experts spend relatively more of their time on understanding the problem, that is, on analyzing the structure of the problem, developing a coherent representation of the problem, and enriching that representation with relevant information retrieved from long-term memory (Simon and Simon, 1978; Voss, et al., 1983). Because experts construct better developed representations and have stored in memory effective procedures for responding in the face of familiar patterns (Gobet and Simon, 1996), they have been found to adopt a working-forward strategy for solving certain problems. Thus, in certain cases experts proceed from the information given, to inferences based on that information, to further inferences, and so on until the goal is reached. Novices, in contrast, often proceed backward from the goal to an equation to calculate that goal, to a second equation to calculate an unknown quantity in the first equation, etc., until an equation is found for which all the quantities needed are part of the given information of the problem. Such a difference between experts and novices has been observed repeatedly in physics, and one small-scale study involving genetics problems in biology adds to the support for this emerging consensus (Smith and Good, 1984).

Students' working backward strategy (referred to as a means-ends analysis in the cognitive science literature), although often effective for solving problems, places a heavy load on working memory, leaving little capacity for learning more general information from the solution attempt, such as a general schema for solving such problems. Working memory refers to the information processing resource that allows a person to (a) hold information in mind temporarily while completing a task or solving a problem and (b) do the work of problem solving (reasoning, language comprehension, etc.). This information burden, known as the working memory load, can be taxing because the working memory system is limited in its capacity to store information and engage in cognitive work (Baddeley, 2007). Thus, it is easy to forget one or more crucial elements of a problem.

As noted by Sweller (1988), when using means-ends analysis, "a problem solver must simultaneously consider the current problem state, the goal state, the relation between the current problem state and the goal state, the relations between problem-solving operators and lastly, if subgoals have been used, a goal stack must be maintained" (p. 261). Solving a problem with a nonspecific goal (e.g., to calculate the value of as many variables as possible) obviates the need to keep several of the aforementioned items in working memory. Indeed, computer simulation work by Sweller (1988) in kinematics, geometry, and trigonometry has demonstrated that problems with nonspecific goals (i.e., open-ended problems) reduce working memory load.

## Physics

Research in physics provides support for the working forward/working backward finding. This research has shown that expert problem solvers typically begin by describing problem information qualitatively and using that information to decide on a solution strategy before writing down equations (Bagno and Eylon, 1997; Chi, Glaser, and Rees, 1982; Eylon and Reif, 1984; Larkin, 1979, 1981a, 1981b; Larkin et al., 1980). A successful solver's strategy includes the appropriate physics concept or principle and, usually, a plan for applying the principle to the particular conditions in the stated problem (Finegold and Mass, 1985; Larkin et al., 1980). This plan leads experts to work forward from the given information to the desired solution. Experts also monitor their progress while solving problems and evaluate the reasonableness of the answer (Chi, 2006; Chi et al. 1989; Larkin, 1981b; Reif and Heller, 1982; Singh, Granville, and Dika, 2002; see Chapter 6 for a more detailed discussion of metacognition). In contrast, beginning physics students typically start by writing down equations that match given or desired quantities in the problem statement and then work backward, somewhat less efficiently, to find an equation for which the unknowns are given directly in the problem (Larkin et al., 1980). When beginning students get stuck using this approach, they lack strategies to go further (Reif and Heller, 1982).[2]

To illustrate the working forward/working backward contrast, consider a problem for which the goal is to determine the final velocity of a block when it reaches the bottom of an inclined plane. As discussed by Larkin, (1981b), expert physicists begin by noting that the motion of the block on the inclined plane depends on gravitational and frictional forces. This approach leads them to retrieve from memory the equation F = ma (force = mass × acceleration). That equation, in turn, leads to retrieval of an equation relating final velocity, the goal of the problem, to acceleration. Novices,

---

[2]This section draws heavily on a review of physics education research that the committee commissioned for this study (Docktor and Mestre, 2011).

in contrast, begin by focusing on the goal of determining the final velocity. This focus leads them to first find an equation that involves that unknown quantity, in this case the equation relating final velocity to acceleration. Acceleration is an unknown in that equation, so novices then look for a new equation that relates acceleration to information given in the problem, in this case $F = ma$.

## Chemistry

Some scholars in chemistry focus not on expert-novice comparisons, but on general problem-solving approaches and on identifying the characteristics of successful problem solvers. According to Herron and Greenbowe (1986), successful problem solvers have a strong command of basic facts and principles; construct appropriate representations; have general reasoning strategies that permit logical connections among the different elements of the problem, and apply verification strategies at multiple points during the problem-solving process. However, similar to findings from physics and cognitive science, research on problem solving in stoichiometry and equilibrium indicates that students are sometimes able to solve a problem using algorithmic/algebraic strategies or analogous problems, with only a superficial understanding of the underlying concept (Camacho and Good, 1989; Chandrasegaran et al., 2009; Gabel and Bunce, 1994; Tingle and Good, 1990). Similarly, a limited amount of research on how students approach organic chemistry problems that involve the use of representations but no calculations suggests that many students memorize the relevant reaction and apply it to a novel task, rather than applying more general skills they have been taught to solve a novel problem (Bhattacharyya and Bodner, 2005).

Some research in chemistry has explored the dichotomy between algorithmic problem solving and problem solving that involves only conceptual understanding (e.g., a multiple-choice question measuring conceptual understanding of gases using no mathematics) (Nurrenbern and Pickering, 1987). A spate of papers in the 1990s probed this subject (e.g., Nakhleh and Mitchell, 1993; Sawrey, 1990), but much of that research used questions that centered on visualizations of the particulate nature of matter (PNOM) as alternatives to algorithmic problem solving. At that time, most textbooks did not contain PNOM problems; because such problems were unfamiliar to students (and many faculty), inferences cannot be readily drawn from those studies. Although PNOM problems are now more common, there is no clear evidence demonstrating whether the use of particulate representations leads to improvements in conceptual problem solving or in problem solving using algorithmic calculations.

*Engineering*

Similar to the cognitive science findings presented above, a limited amount of engineering education research shows that translating the problem into a visual representation and then into a mathematical representation is an important step in solving problems (Eastman, 2001). Even so, students often go straight to a mathematical formula without creating a visual representation of the problem. This approach usually results in failure or misapplication of a formula leading to a dead-end rather than a deeper understanding of the phenomenon under study. One obstacle is students' lack of understanding of concepts that serve as gatekeepers to more sophisticated conceptions of a field (Baillie, Goodhew, and Skryabina, 2006; Meyer and Land, 2005). These concepts are difficult for most students, often abstract, and not recognized by students as keys to new ways of thinking about the discipline. For engineering students, the difficulty also can manifest itself as dependence on ritualistic algorithmic problem solving rather than true understanding, sometimes resulting in the inability of the student to even recognize the problem.[3]

*Summary*

Taken together, these findings suggest that it is important for science and engineering instructors to help students understand that both a good representation of the problem at hand and a good solution method are needed for successful problem solving. Moreover, when students encounter difficulty in solving problems, they need to learn to consider alternative procedures for figuring out the answer and alternative representations of the problem itself (or at least refinements to their current representation). Perhaps one component of an effective instructional strategy would be to provide a compelling example of how much difference a good representation can make for the ease of solution (Posner, 1973, provides one such example).

## Problem Representation

Another consistent finding from DBER and cognitive science is that superficial characteristics of problems have an undue influence on novices' problem solving. One source of cognitive science evidence for this claim is that isomorphic problems (problems that have the same underlying structure) may lead people to construct very different representations of their

---

[3]This section draws heavily from a review of the literature that the committee commissioned for this study (Svinicki, 2011).

(identical) underlying structure because of the different situations they present (e.g., discs of different sizes stacked on a peg versus acrobats of different sizes standing on one another's shoulders), with clear consequences for solution time, accuracy, and method of solving (Bassok and Novick, 2012).

These findings mean that although experts in a domain may easily see that two apparently different problems are really the same kind of problem "deep down," students are likely to assume these problems are of distinctly different types. This assumption impairs students' ability to apply what they learned on one problem to new problems that have similar underlying structures despite superficial differences. Although the initial research in this area involved brain-teaser-type puzzle problems (Hayes and Simon, 1977; Kotovksy, Hayes, and Simon, 1985), similar results have been found more recently for problems from academic domains such as mathematical word problems (Bassok and Olseth, 1995; Bassok, Chase, and Martin, 1998; Martin and Bassok, 2005).

A second source of evidence comes from a large number of studies, in cognitive science and DBER, showing that (relative) experts and novices in a domain (or even good and poor students) differ with respect to the problem features they highlight in their representations. In particular, novices often focus on superficial features of problems, such as the specific objects and terms mentioned and the particular way the question happens to be phrased. Experts, in contrast, typically focus on underlying structural features concerning the relations among the elements in the problem. The structural features are critical for solving the problem and the surface features are not. The following sections on physics, chemistry, and biology education research present results from problem-solving studies that are consistent with this finding. Additional supporting evidence exists based on the results of memory tasks using stimuli from engineering (electric circuit diagrams: Egan and Schwartz, 1979) and biology (clinical cases in medicine: Coughlin and Patel, 1987). Indeed, using a variety of experimental tasks, this finding of expertise differences in problem representations has been replicated in numerous domains, including chess (Gobet and Simon, 1996), computer programming (McKeithen et al., 1981), mathematics (Schoenfeld and Herrmann, 1982), and sports such as basketball and field hockey (Allard and Starkes, 1991). Clearly, this difference is a fundamental aspect of human cognition, relevant to science and nonscience disciplines alike. Part of acquiring expertise in a domain involves learning to identify the important structural features of that domain, and part of being an expert means seeing problems through the lens of the domain's principles (i.e., its deep structure).

Representational differences between experts and novices have implications for problem solving accuracy, solution time, and the ability to transfer analogous solutions across superficially dissimilar problems (Hardiman,

Dufresne, and Mestre, 1989; Novick, 1988; Novick and Sherman, 2008). This coupling of representation and solution occurs because the perceptual configurations and structural relationships that experts discern are often associated with stored plans in long-term memory concerning what to do in the presence of those configurations. Thus, experts often see solutions that novices have to laboriously compute (Chase and Simon, 1973).

## Physics

Physics education research has revealed differences in how experts and novices represent problems. Novice representations include physical objects (e.g., a drawing of a car in a problem with acceleration), whereas scientific or expert representations add details based on the laws of physics and incorporate abstractions (e.g., a graph of the car's motion) (Larkin, 1983). A common task used to investigate the nature of people's problem representations involves having them categorize problems according to how they are solved. Physics education research has found that whereas experts categorize physics problems according to the major concepts or principles that can be applied to solve them (e.g., Newton's second law), novices rely much more on the surface attributes of the problems—such as the specific objects mentioned (e.g., pulleys versus inclined planes versus springs)—as the basis for categorization (Chi, Feltovich, and Glaser, 1981; de Jong and Ferguson-Hessler, 1986). Nevertheless, depending on the problem description, even experts sometimes have difficulty focusing on the major principle in certain types of categorization tasks, and some novices are able to consistently rely on principles to make categorization decisions (Hardiman, Dufresne, and Mestre, 1989). Across different levels of expertise, physics principles play a key role in the organization of conceptual and procedural knowledge for those who are good problem solvers.

Expertise-related differences in problem representations have important implications for teaching and learning because naïve representations may incline novices to develop incorrect assumptions based on common sense (Anzai, 1991). On the other hand, an instructional intervention in which physics students were directed to attend to the underlying structure of problems in the way that experts do yielded more expert judgments of the extent to which problems were similar, which resulted in improved problem solving (Dufresne et al., 1992).

## Chemistry

Problem representation has not been a significant focus of research on chemistry problem solving. However, a limited amount of small-scale research does support the finding that novices focus on superficial aspects of problems. In a study of the mental models that organic chemistry students use when making predictions about the relative acid strength of different substances, 7 of 19 students interviewed had mental models that caused them to classify acids and bases by surface features and simple heuristics rather than using information about the structures themselves to predict and explain the properties of those structures (McClary and Talanquer, 2010). Another interview study of how students use their understanding of hydrogen bonding to characterize molecules revealed that after four semesters of chemistry, students were still unable to easily identify concepts that were relevant—or not relevant—to the problems they were solving (Henderleiter et al., 2001). In addition to documenting the intransigence of students' difficulties in identifying the critical attributes of a problem, a few studies have linked these difficulties to challenges of transferring existing knowledge to a new concept (Kelley and Jones, 2008; Tien, Teichert, and Rickey, 2007; see Chapter 7 for a more detailed discussion).

*Biology*

Two small-scale interview studies from biology also illustrate expert-novice differences in problem representation. Smith (1992) found that undergraduate biology students grouped classical genetics problems based on superficial features (e.g., whether the problem concerned humans or fruit flies, how the question was worded), whereas biology professors grouped them according to key underlying concepts (e.g., the mechanism of inheritance). Genetics counselors showed an intermediate pattern of grouping. In another study, Kindfield (1993/1994) investigated the diagrams drawn by 15 participants with more versus less accurate knowledge of meiosis and chromosomes (three professors, two graduate students, and four undergraduate honors students in genetics compared with one honors student and four biology majors enrolled in an introductory genetics course) while solving problems involving meiosis. Kindfield found that less knowledgeable participants often drew chromosome representations that more literally resembled chromosome appearance under a light microscope, and included features that are irrelevant to solution (e.g., dimensionality and shape). In contrast, the drawings of more knowledgeable participants included chromosome features that are biologically relevant to the given problem.

## Individual and Group Differences

Across DBER, individual and group differences are not a common focus of study. However, some research in chemistry and biology has investigated group differences in problem solving. Although some of this research has been conducted on a relatively large scale (i.e., with more than 100 students), the committee has characterized the strength of conclusions that can be drawn from this research as limited because few studies exist.

## Chemistry

Most chemistry education research on differences in problem solving has examined differences in cognitive abilities rather than demographic characteristics. One example relates to working memory capacity. By their very nature, complex problems contain a considerable amount of relevant information, which often must be held in students' working memory. Some research on problem solving in chemistry indicates that as the working memory load of a problem increases, success rate decreases (Johnstone and El-Banna, 1986; Niaz, 1989). That research also has shown that chemistry students who have a larger working memory capacity (as measured by the Pascual-Leone FIT test) perform better on problem solving. This research has been conducted on a relatively large scale and the findings are consistent with a large body of research in cognitive psychology documenting enhanced performance (e.g., language comprehension, problem solving, etc.) when more working memory resources are available to complete the task at hand (e.g., Baddeley, 2007).

Other research on individual differences has examined the relationship between spatial thinking, or the ability to mentally manipulate two- and three-dimensional figures, and problem solving in chemistry. Here, the evidence is mixed. On the one hand, Bodner and McMillan (1986) have found that problem solving activities appear to be correlated with spatial skills. Stieff, Hegarty, and Dixon (2010) also have reported evidence for a correlation between spatial skills and molecular visualization (or interpreting visual representations of molecules), which may affect performance on some kinds of chemistry problems. On the other hand, Wu and Shah (2004) have postulated that other cognitive factors may be correlated with spatial skills, which may explain Bodner and McMillan's results.

## Biology

In contrast to chemistry, biology education research on group differences has concentrated on demographic characteristics. A series of studies used the Interactive MultiMedia Exercises (IMMEX), a multimedia software program that allows educators to track students' progress as they solve problems (Stevens and Palacia-Cayetano, 2003). Comparisons of the performance of two-year college and university students on numerous IMMEX exercises revealed several differences between the two groups. In particular, community college students more frequently used simple statements that repeated text from the exercise and asked fewer questions during decision making, which indicated a lower degree of analysis and hypothesis forming. On the other hand, community college students showed greater awareness of their progress (or lack thereof) toward the task goals than university students, who were more successful in problem solving overall.

## Ill-Defined Problems

Relatively little DBER has been conducted on ill-defined, open-ended problems. Most of this research comes from chemistry. Some engineering education research also investigates ill-defined problems, typically in the context of instructional strategies (discussed under "Instructional Practices to Improve Problem Solving"). The committee has characterized the strength of the conclusions that can be drawn from this evidence as moderate because although relatively few studies exist, they have been conducted on a relatively large scale (i.e., across multiple courses or institutions, with several hundred students) and include quasi-experimental studies.

### Chemistry

As mentioned, IMMEX is a problem-solving assessment system. Because many IMMEX assignments are case-based problems, some research in chemistry has used IMMEX to investigate students' solving of ill-defined problems. Some research on a qualitative analysis problem requiring students to identify an unknown compound based on the results of the physical and chemical tests that they request in the context of the IMMEX problem has shown that after five attempts at solving a problem, students stabilize on a problem-solving strategy, even when that strategy is unsuccessful (Cooper et al., 2008).

Other studies on ill-defined problems have used the IMMEX system to measure the effects of group problem-solving sessions or an intervention designed to increase metacognitive activity (Sandi-Urena, Cooper, and Stevens, 2011). The results suggest that students working on ill-structured problems in collaborative groups are better able to describe their problem-solving strategies than students working alone. Also, when solving problems individually immediately after the grouping, students who had worked in a group retained the strategies that the group developed (Cooper et al., 2008).

### Instructional Practices to Improve Problem Solving

Across all disciplines, instructional strategies to improve students' conceptual understanding, problem solving, and overall academic performance are the subject of considerable inquiry in DBER. In this section, we highlight research that specifically investigates strategies to improve problem solving (see Chapter 4 for a discussion of research on strategies for promoting conceptual change and Chapter 6 for a discussion of research on more general instructional strategies).

## Physics

Whereas early research in physics identified problem-solving differences between experts and novices, current research on physics problem solving primarily investigates how to move students from novice toward expert problem solving within the domain. To this end, DBER has investigated the efficacy of various types of support for students (or scaffolding), including the following:

- using a specific problem-solving framework (Heller and Heller, 2000; Pólya, 1945; Reif, 1995; Van Heuvelen, 1991),
- elucidating different problem types (Mestre, 2002; Van Heuvelen, 1995; Van Heuvelen and Maloney, 1999),
- providing example solutions (Chi et al., 1989; Ward and Sweller, 1990), and
- altering the classroom format to provide more guidance and interaction (Cummings et al., 1999; Duch, 1997; Hoellwarth, Moelter, and Knight, 2005).

Other areas of inquiry include the efficacy and use of example problems (Cohen et al., 2008), nontraditional problem types (Ogilvie, 2009), computer coaches (Gertner and VanLehn, 2000; Hsu and Heller, 2009; Reif and Scott, 1999), and cooperative group interactions (Heller and Hollabaugh, 1992; Heller, Keith, and Anderson, 1992). This research is being conducted in a wide range of settings from small-scale research laboratory situations to large classroom studies, and the committee has characterized the strength of findings from the research on this topic as strong.

Overall, this research indicates that expert skills in physics problem solving can be taught and that carefully designed support appears to be beneficial for students. However, individual studies suggest that problem-solving gains from a given type of scaffolding are small and difficult to measure. Moreover, various types of scaffolding are interrelated, which makes individual effects difficult to determine.

Another line of physics education research has shown that making symbols more transparent to students helps them apply concepts and solve problems (Brookes and Etkina, 2007). For example, instead of writing forces as W (weight) or T (tension), students benefit when labeling each force with two subscripts to identify two interacting objects, such as $F_{EonO}$ (force exerted by Earth on object) or $F_{RonO}$ (force exerted by rope on object). As another example, to help students understand that heat is a process of energy transfer and not energy itself, the term heat can be substituted with "heating." Although some physics education research-based curriculum materials use this more descriptive language (Van Heuvelen and Etkina, 2006), research on the efficacy of those materials is scant.

These results suggest that a systematic approach to improving such a complex skill as problem solving, with multiple forms of supporting scaffolding, is indicated. Such approaches typically are consistent with the theoretical framework of cognitive apprenticeship (Brown, Collins, and Duguid, 1989; Yerushalmi et al., 2007), which proposes that complex skills depend on an interlocking set of experiences and instruction whose efficacy, in turn, depends on the learner and the community of practitioners with whom the learner interacts (see Box 5-1 for a discussion of cognitive apprenticeship).

### Chemistry

Teaching general problem-solving methods is a significant emphasis of chemistry education research. Pólya (1945) recommended teaching an organized approach to solving problems in mathematics, and developed the strategy of (1) understanding the problem, (2) devising a plan, (3) carrying out the plan, and (4) looking back. Other scholars have modified that approach for chemistry and studied the effects of using those approaches on problem solving. A summary of this literature concluded that "when chemistry students are taught to solve problems in a systematic manner, they are more successful. Strategies based on Pólya's heuristics or variations thereof appear to facilitate students' ability to solve routine problems even though there is some evidence that students may be doing so using algorithms" (Gabel and Bunce, 1994, p. 318). Despite these benefits, some research suggests that many students do not continue using the strategies they were initially taught (Bunce and Heikkinen, 1986).

### Engineering

Engineering education effectively has incorporated key elements of cognitive science into problem-solving and design[4] experiences for engineering courses and curricula, and into the research about effects on learning (e.g.,

---

[4] "Design" has distinct meanings across engineering sub-disciplines. In mechanical and civil engineering, design typically has a physical connotation. In computer science and computer engineering, design can be different both in the scale of application and the use of simulation instead of actual construction (Dutson et al., 1997).

---

**BOX 5-1**
**Cognitive Apprenticeship**

A chemistry postdoctoral researcher shows an undergraduate how the gas chromatograph/mass spectrometer needs to be calibrated; a young astronomer watches her mentor give a research talk at a conference; an engineering student uses a computer-aided design program for the first time and follows step-by-step instructions prepared by his professor. In each of these cases, skills—and often cultural norms—are passed down from an expert to a novice. History recognizes this process as apprenticeship.

Traditional apprenticeship includes four dimensions:

1. Modeling, where the expert demonstrates how to do a task.
2. Scaffolding the various types of support that the expert provides as the novice proceeds.
3. Fading, or the gradual removal of scaffolding as the novice takes more responsibility.
4. Coaching by the expert to guide the novice toward success, for example by pointing out strengths and weaknesses in the novices' product or performance.

Cognitive apprenticeship builds on the traditional apprenticeship model, but makes every aspect more explicit and relevant. Developed by Collins, Brown, and Newman (1989), cognitive apprenticeship has been shown to be successful in teaching mathematical problem solving, reading, and writing. Cognitive apprenticeship has been used as a framework to explore the affective side of undergraduate research experiences (Sadler et al., 2010; see Chapter 7), develop physics problem solving (Heller, Keith, and Anderson, 1992), and develop introductory engineering software skills (Hundhausen et al., 2011). Although cognitive apprenticeship is not useful for teaching rote knowledge, it is a robust and well-researched technique for teaching complex tasks in many disciplines.

---

Atman, Kilgore, and McKenna, 2008; Cordray, Harris, and Klein, 2009). The committee has characterized the strength of conclusions that can be drawn from this research as moderate because the body of research includes a review and a line of research with multiple studies that were conducted in multiple courses.

Particularly in design, it is important to use authentic problems and to sequence experiences within various courses to support the learning of core concepts. Strategies such as case analyses, model-eliciting activities, worked-out problem examples, and heuristics provide opportunities for students to gain experience and confidence in solving open-ended problems (Sankar, Varma, and Raju, 2008; Svinicki, 2011). Indeed, a series of mixed-methods studies in engineering has shown that model-eliciting activities in which teams of students work with open-ended, real-world problems can improve students' problem solving (Yildirim, Shuman, and Besterfield-Sacre, 2010). However, these activities must be implemented correctly to be effective. Some factors that contribute to effective implementation of model-eliciting activities include the nature and quality of guidance and feedback provided by the instructor, duration of the activity, size of student teams, and instructors' level of experience with model-eliciting activities (Yildirim, Shuman, and Besterfield-Sacre, 2010). Incorporating reflection and self-explanation prompts into instruction also has been shown to improve student problem solving (Cheville, 2010; Cordray, Harris, and Klein, 2009; Litzinger et al., 2010; Svinicki, 2011) (see Chapter 6 for a more detailed discussion of metacognition).

*Biology*

The strength of conclusions that can be drawn from biology research on problem solving is limited because the research base consists of a few studies that have been conducted in the context of single courses. However, consistent with findings from physics and engineering, the results suggest that problem-solving skills can be enhanced through instruction. One quasi-experimental study examined the use of "invention activities" as a way to improve biology problem-solving strategies (Taylor et al., 2010). Invention activities are based on work that challenges students to solve problems that seem unrelated to current class material, which then helps students to construct a mental framework that promotes better understanding of the course content. Students in a first-year biology class who participated in an invention activity began working on problems much more quickly and generated more hypotheses (many of which were plausible) than students who did not participate in the invention activity (Taylor et al., 2010).

As discussed in Chapter 6, collaborative problem-solving activities are increasingly popular in physics, chemistry, and biology. Some recent work has been done to develop and validate tools for comparing collaborative and individual problem-solving strategies in large (60-100 students) biochemistry courses where students discuss ill-defined problems in small online groups (Anderson et al., 2008) and then work on individual exams based on similar, but not identical, problems (Anderson et al., 2011). Both of these assessment tools allow students to regularly practice their problem-solving skills, and, perhaps more importantly, allow instructors the opportunity to provide targeted intervention, as appropriate, to groups and individual students in this critical area of student development. Studies using these instruments are ongoing and have not yet been published.

## The Geosciences

Problem-solving skills in the geosciences are commonly taught in the context of problem-based learning (Macdonald et al., 2005), an instructional approach in which students are given complex, realistic problems to solve individually or in groups (Barrows, 1986). With instructional strategies and activities that are influenced by Bransford, Vye, and Bateman (2002) and Kolb (1984), problem-based learning in the geosciences often involves ill-defined problems that have applications to society, such as environmental issues, public policy, geology and human health, natural hazards, and Earth resources (see Ishikawa et al., 2011 for an example). The products of these learning activities typically are measured against professional norms of geoscience research projects, such as geologic maps and written reports (Carlson, 1999; Connor, 2009; de Wet et al., 2009; Fuller et al., 2006; Gonzales and Semken, 2009; Maskall and Stokes, 2008; May et al., 2009; Potter, Niemitz, and Sak, 2009).

For the most part, the geoscience education literature describes these activities and does not examine their efficacy. However, Ault (1994) made a notable early effort to formulate a research agenda on problem solving for Earth science education, laying out the difficulties that confront students. These difficulties included incomplete observations, systems that are "complex beyond complete prediction" (Ault, 1994, p. 275), systems with contingent histories in which extrapolation is unreliable, and value-laden contexts surrounding socially critical problems. In addition, as discussed in Chapter 6, technologies such as GPS tracking devices are being used to monitor student navigation and yield insight into problem-solving skills (Riggs, Balliet, and Lieder, 2009; Riggs, Lieder, and Balliet, 2009). That research reported an optimum amount of relocation and backtracking in field geology: too much retracing indicates confusion, and too little reoccupation of key areas appears to accompany a failure to recognize important geologic features.

### Summary of Key Findings on Problem Solving

- *The different fields of discipline-based education research have studied somewhat different aspects of problem solving, consistent with disciplinary differences in what problem solving entails. Early research in physics emphasized expert-novice differences in various aspects of the solution process, whereas chemistry education research and later research in physics has examined the nature of the solution process and strategies for successful problem solving more broadly. Most of the research in engineering has addressed instructional strategies, particularly around authentic problems; instruction is also an important area of inquiry for physics. Research on problem solving in biology and the geosciences examines individuals' problem-solving processes, rather than expert-novice differences.*

- *With the exception of engineering and chemistry, discipline-based education research on problem solving typically focuses on well-defined, quantitative problems. In most disciplines, considerably less research exists on problems that are more characteristic of what scientists and engineers encounter in their professional lives.*

- *Part of acquiring expertise in a domain involves learning to identify the important structural features of that domain, and part of being an expert means seeing problems through the lens of the domain's principles. Students (novices) have difficulty with all aspects of problem solving and approach well-defined problem solving in ways that are consistently and identifiably different from those of experts. Specifically, experts often preferentially focus on creating a representation of the problem at the outset. In addition, experts and undergraduate students represent problems in very different ways, with experts attending to the underlying principles required for solution and students focusing on superficial features such as the particular objects mentioned in the problems. These differences have important implications for problem solving success and, in turn, instruction.*

- *As cognitive science research has shown more broadly, insufficient spatial skills and working memory capacity may impede performance on some forms of problem solving.*

- *Several conditions and strategies appear to improve problem-solving skills, including socially mediated learning environments, the use of open-ended and authentic problems, and interventions that promote metacognitive activity. Nonetheless, some research shows that students often do not persist in using effective strategies over the long term.*

### Directions for Future Research on Problem Solving

Problem solving has been a significant focus of DBER in many disciplines, and DBER has generated important insights into the nature of the problem-solving process, differences between experts and novices, and strategies for improving problem solving. However, some areas remain ripe for exploration. For example, little DBER has investigated individual and group differences in problem solving. A better understanding of similarities and differences in problem solving among subpopulations of students is needed, along with careful studies investigating the effects of various individual differences (spatial skills, working memory capacity, logical thinking ability) and their implications for instruction.

In all disciplines, DBER scholars have a pressing need for measurement tools that will assess student problem-solving skills for large numbers of students in an authentic classroom setting. Some of these tools have been developed in physics (Docktor and Heller, 2009), and in chemistry (Cooper et al., 2008; Sandi-Urena and Cooper, 2009), but they have yet to be applied widely enough to gauge their utility.

The committee's analysis revealed that most DBER on problem solving is aimed at identifying, categorizing, and clarifying what students do wrong. Although this is an important first step, more resources should now be devoted to investigating the effect of instructional interventions on students' problem-solving skills (similar to Mestre and Ross, 2011, for example, or Sandi-Urena and Cooper, 2011). Many research papers offer implications for teaching, but relatively few actually investigate the effect of an intervention designed to improve problem solving in some context (most of these studies come from physics, chemistry, and engineering). A productive approach might be to identify trajectories that can lead to greater problem-solving expertise (Lajoie, 2003). Further research on attributes that might characterize trajectories toward problem-solving competence could include data on interpersonal interactions when discussing problems, and the use of methodologies such as eye tracking (Smith, Mestre, and Ross, 2010).

Systematic explorations also are needed of the effects of changing problem features (e.g., format of the problem statement; familiarity with the problem context; whether the values provided are numeric or symbolic) and the effects of different types of worked examples on problem-solving performance. In addition, strategies should be developed for effectively reducing working memory load while still highlighting important aspects of problem solving. Sweller and colleagues (Owen and Sweller, 1985; Sweller and Levine, 1982; Sweller, Mawer, and Ward, 1983) in psychology have shown that one effective way to accomplish this goal is to have students engage in open-ended problem solving rather than attempt to reach a particular goal (e.g., "Calculate the value of as many variables as you can" versus "What is the final velocity of the car?"). As discussed, the open-ended approach reduces the number of variables and problem conditions that students must keep in their working memory. This research has used problems in kinematics, geometry, and trigonometry, which suggests that the technique may be widely applicable to the science and engineering domains that are the purview of this report.

Real-world problems are often considerably different from the types of problems students typically solve in the laboratory or class, because they are ill-defined, messy, and knowledge intensive (Novick and Bassok, 2005). Most DBER on problem solving addresses well-defined problems. Substantial work remains to be done to understand how students approach less well-defined problems and how to improve their ability to tackle these problems. Moreover, discipline-based education researchers must investigate the extent to which instructional strategies for improving students' problem solving for well-defined and real-world problems transfer to ill-defined problems as well. Although well-defined numerical problems are typical in traditional science and engineering courses, most practicing scientists and engineers typically solve conceptual problems, such as deducing structure with spectroscopic tools and designing experiments or approaches to target molecules. One strategy to address this complex issue would be to compare students' performance on problems at differing levels of approximation to real-world problem solving.

The consideration of well-defined versus ill-defined problems naturally leads to the distinction between problem solving and problem finding (or problem discovery). Before professionals engage in problem solving, they must confront the critical initial step of choosing which problem(s) to solve. Not all problems are equally worthy of extended effort, and not all are equally likely to result in significant insights or breakthroughs. For scientists and engineers, therefore, problem finding is crucial. In most educational situations, however, students are simply provided problems by their teachers. Instructors and discipline-based education researchers may underemphasize the importance of problem finding because, traditionally at least, it is not a task in which students are expected to engage. Thus, research is needed on factors that affect students' ability to engage in problem finding.

In biology and the geosciences, rich opportunities exist to conduct research on ill-defined or open-ended problems. As physics education research and chemistry education research have done, this work can readily extend the more general research on the development of problem-solving skills in cognitive science. In the geosciences, it would be productive to build on emerging research on cognitive apprenticeship and related work on metacognition in which geoscientists reveal their thought processes in a mentoring capacity to students (Manduca and Kastens, 2012; Petcovic and Libarkin, 2007).

## THE USE OF REPRESENTATIONS AND SPATIAL THINKING IN PROMOTING CONCEPTUAL UNDERSTANDING AND PROBLEM SOLVING

Visual/spatial, mathematical, logical, and verbal representations are central to human thinking, as well as to learning and instruction in virtually all disciplines. Considering only visual/spatial representations, McKim (1980, p. 142) noted that "...you will find graphic-language expressions on the blackboards of almost every department of a university, from aeronautics to zoology."

Representations like diagrams, graphs, and mathematical equations are important in science and engineering because they facilitate communication, aid in the discovery of scientific facts, assist in problem solving, serve as memory aids, and generally function as tools for thinking (Dufour-Janvier, Bednarz, and Belanger, 1987; Kindfield, 1993/1994; Larkin and Simon, 1987; Lynch, 1990; Novick, 2001). In addition, abstract spatial representations can actually modify, and in some cases simplify, the nature of a task (Larkin and Simon, 1987). For example, people can usually estimate proportions more easily from a pie chart than from a table of numeric data. In addition to interpreting representations produced by experts, students can construct their own representations to enhance learning and engagement (Ainsworth, Prain, and Tytler, 2011).

Because a wide variety of visual/spatial representations play a central role in science and engineering, students need to develop skills in constructing, interpreting, transforming, and coordinating these representations (see Box 5-2 for a description of the obligatory components of representations that students must understand). This suite of skills has been termed representational competence (Hegarty, 2011).

Developing representational competence requires the ability to mentally manipulate two- and three-dimensional objects, a skill that is called visuospatial thinking or spatial ability. Given the importance of spatial representations and thinking processes in science, it seems logical that spatial ability is correlated with participation in science generally (Shea, Lubinski, and Benbow, 2001; Wai, Lubinski, and Benbow, 2009; Webb, Lubinski, and Benbow, 2007). However, as the following discussion illustrates, evidence from DBER on the role of spatial ability is mixed.

## Overview of Discipline-Based Education Research on the Use of Representations and Spatial Thinking

Spatial thinking and the use of representations are important areas of inquiry for DBER because representational competence is so vital to acquiring expertise within a discipline. For example, the visualization and representation of the unseen molecular level is central to a robust understanding of chemistry. Chemistry students must learn how to create novel, discipline-specific representations, how to translate those representations into the more familiar equation format, and how and when to apply each representational format for solving problems. As another example, the ability to visualize a three-dimensional image from a two-dimensional graphical representation,

---

**BOX 5-2**
**The Four Components of Representations**

Representations have four obligatory components, described by Markman (1999):

1. The situation that is being represented (called the "represented world"). In the science and engineering disciplines, the represented world might be a problem to be solved or a device to be designed.
2. The elements used to construct the representation (called the "representing world"). These elements could be any of the following:
   (a) physical objects, such as when balls and sticks are used to model the structure of a chemical compound;
   (b) diagrammatic elements such as lines and geometric figures that can be used, for example, to create a sketch of the internal structure of a cell or a graph showing the acceleration of an object over time;
   (c) mental elements, such as components of a mental model of a pulley system that students may attempt to animate mentally to reason about the behavior of the system.
3. Rules for mapping elements of one world to those of the other world, or a set of conventions for constructing a particular type of representation (Hegarty, Carpenter, and Just, 1991), which students must learn if they are to successfully use appropriate representations to solve problems and reason about scientific phenomena.
4. Processes that operate on the representing world. One set of processes uses the mapping rules to create the representation. Other processes then extract information from the representation to reason about the represented world.

---

and the ability to mentally rotate three-dimensional objects, are important to success in engineering, and especially so in engineering design (Sorby, 2009; see Box 5-3 for a description of representations used in design).

DBER on representations and spatial ability is particularly valuable in understanding the many disciplinary differences that pertain to thinking visually and understanding visual representations. For example, in some disciplines (e.g., geology, anatomy), penetrative thinking, or the ability to represent and reason about the hidden internal structure of a multilayered three-dimensional object, is critically important (Hegarty, 2011; see "The Role of Spatial Ability in Visualization and Mental Model Formation" in

---

**BOX 5-3**
**Representations in Engineering: The Languages of Design**

Representations are an essential component of engineering design, and the following representations constitute a major part of the "languages of design" (Dym et al., 2005, p. 108):

*verbal or textual statements* used to articulate design projects, describe objects, describe constraints or limitations, communicate between different members of design and manufacturing teams, and document completed designs;

*graphical representations* used to provide pictorial descriptions of designed artifacts such as sketches, renderings, and engineering drawings;

*shape grammars* (a computational approach to the generation of designs) used to provide formal rules of syntax for combining simpler shapes into more complex shapes;

*features* used to aggregate and specialize given geometrical shapes that are often identified with particular functions;

*mathematical or analytical models* used to express some aspect of an artifact's function or behavior, where this behavior is in turn often derived from some physical principle(s); and

*numbers* used to represent discrete-valued design information (e.g., part dimensions) and parameters in design calculations or within algorithms representing a mathematical model.

---

this chapter). Mental animation ability may be more important in disciplines for which motion is a central concern (e.g., mechanics, meteorology, oceanography).

Disciplines may also differ in the extent to which they use graphical versus mathematical representations as the means of conveying critical insights. In a comparison of representations in scientific papers across several scientific fields, authors of geoscience and chemistry papers were found to use the most figures whereas authors of physics papers inclined more toward equations and psychology authors toward tables (Kastens and Manduca, 2012). Biology is increasingly coming to rely on mathematical representations, especially to interpret large genomic databases. In chemistry, spatial and mathematical representations are both important.

Different science and engineering disciplines also may differ in the extent to which they call upon large-scale versus small-scale spatial ability (Hegarty et al., 2010). Small-scale spatial ability refers to one's skill at perceiving and imagining transformations of objects that can be manipulated. In contrast, large-scale (or environmental) spatial ability refers to one's success in tasks such as learning the layout of a new environment, retracing a previously traveled route, and pointing to unseen locations in a familiar environment. The two kinds of spatial information may be processed separately by the human visual system (Previc, 1998). Hegarty et al.'s (2010) self-report data from scientists, engineers, and others suggest that large-scale spatial ability may be particularly important in the geosciences, whereas small-scale spatial ability may be particularly important in engineering. It should be noted, however, that tools for assessing small-scale spatial abilities are far more advanced than those for assessing large-scale spatial abilities (Hegarty, 2011).

This section discusses DBER on how students develop, use, and interpret representations, and the role of spatial thinking in visualization and mental model formation. Because representations are so important to human thinking and learning, a large literature in cognitive science addresses students' understanding of visual/spatial, mathematical, logical, and verbal representations (see, for example, Glasgow, Narayanan, and Chandrasekaran, 1995; Thagard, 2005). A critical theme in this research is the importance of general cognitive and perceptual factors for understanding and reasoning with representations in science and engineering, as well as features related to the diagrams themselves. Because the breadth of cognitive science research in this area reflects the cross-disciplinary importance of representations, much of this research has used materials and tasks from various science disciplines. Given this interdependence between DBER and cognitive science, we use findings and principles from cognitive science to frame the findings of DBER.

*Research Focus*

Many of the disciplinary differences just described are reflected in the focus of DBER. In physics and chemistry, the research base on spatial ability and the use of representations is strong because multiple studies exist—many of relatively large scale—with high convergence of findings. As characterized by Docktor and Mestre (2011, p. 20), physics education research explores the use of external representations for describing information during problem solving, such as pictures, physics-specific descriptions (e.g., free-body diagrams, field line diagrams, or energy bar charts), concept maps, graphs, and equations. Some studies investigate the representations students construct during problem solving and how they use those representations (Heller and Reif, 1984; Larkin, 1983; Rosengrant, Etkina, and Van Heuvelen, 2005; Van Heuvelen and Zou, 2001), whereas other studies explore the facility with which students or experts can translate across multiple representations (Kohl and Finkelstein, 2005; Meltzer, 2005). Still other studies in physics focus on student difficulties with particular representations like graphs or mathematical formulae (Beichner, 1994; Ferrini-Mundy and Graham, 1994; Goldberg and Anderson, 1989; McDermott, Rosenquist, and van Zee, 1987; Orton, 1983).

More than 100 studies exist on spatial thinking and representations in chemistry, including books (Gilbert and Treagust, 2009) and reviews (Taber, 2009). These studies examine difficulties translating between representations (Gilbert and Treagust, 2009; Johnstone, 1991), the role of spatial ability in visualization and mental model formation (Abraham, Varghese, and Tang, 2010; Bodner and McMillan, 1986; Pribyl and Bodner, 1987; Stieff, 2011), and the influence of animated and static visualizations on conceptual understanding (Abraham, Varghese, and Tang, 2010; Aldahmash and Abraham, 2009; Sanger and Bader, 2001).

Spatial ability and the use of representations are emerging areas of study in engineering (Sorby, 2009), biology (Dirks, 2011) and the geosciences (Piburn, van der Hoeven Kraft, and Pacheco, 2011). In engineering, much of this research addresses instructional approaches to improve spatial ability; these approaches typically are grounded in constructivist theories of learning (Gerson et al., 2001). Most of the biology studies investigate the role of different representations in improving understanding, promoting conceptual change, and stimulating interest in biology. Many are grounded in constructivism and dual-coding theory. The latter theoretical approach was refined in the model of Schönborn and Anderson (2009), which suggested that three factors may affect learners' ability to interpret external representations in biology and biochemistry: students' prior knowledge of the concepts underlying the external representation, students' reasoning ability, and the mode or nature of the diagram. In the geosciences, research in this area has concentrated on the relationship between spatial ability and success in the geosciences (Liben, Kastens, and Christensen, 2011) and on instructional strategies to improve spatial ability (Piburn et al., 2005).

*Methods*

DBER scholars employ a wide range of methods to study spatial thinking and representations, including case studies on how students construct and use representations, think-aloud interviews, student-generated diagrams, surveys of students' confidence and knowledge, pre- and post-tests of content knowledge, quasi-experimental studies, large-scale assessments, and mixed methods studies using technology to record and aggregate the data. Study populations range from nonmajors in introductory courses to graduate students, and some studies include high school students (Beichner, 1994; Swenson and Kastens, 2011). Research conducted in the context of physics problem solving typically involves qualitative investigations into the thinking of individual students in introductory physics courses. In many cases, problem solvers are carefully observed while they do their work, and their written solutions are examined to see how they make use of different representations.

Indepth studies of the use of graphical representations in science and engineering necessarily include specific representations and tasks. This focus raises questions about the generalizability of these studies within their disciplines. Moreover, although research from the sciences and engineering is broadly concerned with spatial structures, the disciplines engage with dramatically different scales of time and space, which raises important questions of how spatial representations differ across the various science disciplines and whether students face common challenges across the disciplines.

### How Students Develop, Use, and Interpret Representations

A critical step in helping students acquire greater disciplinary expertise is to understand how they develop, use, and interpret the representations that are central to a given discipline. As this discussion shows, research from physics, chemistry, biology, and the geosciences consistently indicates that students have difficulties interpreting representations, and that they struggle to see similarities among different representations that describe the same set of relationships.

More specifically, research in cognitive science has shown that across disciplines, students have (a) a preference for representations that have a high degree of visual similarity to their referents; (b) a bias to view representations as truthful depictions of reality; and (c) particular difficulty understanding the abstract nature of representations (similar to findings under "Problem Solving" that experts are more able than novices to see beyond the surface of what things look like to capture the essential nature of situations within their domain of expertise) (Hegarty, 2011). Knowing the conventions for how diagrams represent reality is not sufficient to ensure good comprehension or successful inference with these representations. Experts see patterns in representations that novices miss, even when novices know the conventions, because novices are distracted by salient but irrelevant surface features of the displays (see "Problem Solving" for more details).

### Physics

A variety of studies in physics supports the finding that students have serious difficulties simply interpreting the representations commonly encountered in introductory physics courses. These studies often involve microcomputer and video-based labs and simulations on vectors, electric field lines, electric and magnetic fields, or waves (see Rosengrant, Etkina, and Van Heuvelen, 2007 for a summary of the research). Some research indicates that students do not use diagrams because they misunderstand the quantities and concepts being represented, and because they are given few chances to specifically practice the skills needed to construct diagrams (Van Heuvelen, 1991). However, some research—including a quasi-experimental study with approximately 600 students at two universities (Kohl, Rosengrant, and Finkelstein, 2007)—shows that learning via more than one representation can help to build understanding (Van Heuvelen and Zou, 2001; see Box 5-4 for an example).

*Chemistry*

The relationship between structure and properties is central to chemistry, and students are expected to understand this relationship by the time students take organic chemistry and biochemistry. However, as we discussed in Chapter 4, an abundance of research on the macroscopic, symbolic, and particulate domains—three essential domains of chemistry learning that are depicted in Johnstone's triangle (1991, see Box 5-5 and Chapter 4)—has revealed students' difficulties in understanding the particulate nature of matter (the fact that matter is made of discrete particles). Specifically, students have difficulty constructing Lewis structures, commonly used particulate-level diagrams of chemical and physical phenomena that demonstrate molecular shape and bonding (Cooper et al., 2010; Nicoll, 2003). In a study of 166 undergraduate students in general and organic chemistry, the accuracy of students' representations fell precipitously when the number of atoms on the Lewis structure increased from six to seven (Cooper et al., 2010). The increase from six to seven molecules represents a shift to an atomic structure where the bonds are not evident unless students recognize the molecule. That study, along with another study investigating students' understanding of the Lewis structure of the molecule $CH_2O$, also demonstrated that students' understanding of these representations does not improve over time (Cooper et al., 2010; Nicoll, 2003).

Students also have difficulty translating among representations at the other two corners of Johnstone's triangle (Gilbert and Treagust, 2009; Johnstone, 1991; Linenberger and Bretz, 2012). More generally, students have difficulty translating among alternative representations that describe the same set of relationships, such as videos, graphs, animations, equations, and verbal descriptions (Kozma and Russell, 1997). However, as with physics, the use of multiple representations appears to be beneficial. For example, in one quasi-experimental study, female students' performance on problems improved more than males' when students were taught particulate representations in addition to symbolic representations (Bunce and Gabel, 2002).

### BOX 5-4
### Visualization/Representation in Physics

The common free-body diagram helps students move from a physical view of concrete objects to the more abstract idea of vector forces. This is a major component of a multirepresentational teaching methodology developed by Alan Van Heuvelen (1991). Van Heuvelen and Zou (2001) found this approach to be more effective than traditional instruction.

**(a) Words**

A parachutist whose parachute did not open landed in a snow bank and stopped after sinking 1.0 m into the snow. Just before hitting the snow, the person was falling at a speed of 54 m/s. Determine the average force of the snow on the 80-kg person while sinking into the snow.

**(b) Pictorial Representation**

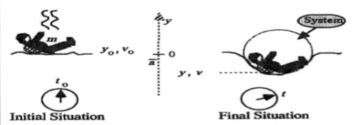

Initial Situation        Final Situation

**(c) Physical Representation**

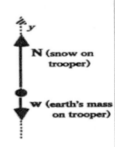

N (snow on trooper)

w (earth's mass on trooper)

**(d) Math Representation**

$$2\bar{a}(y - y_0) = v^2 - v_0^2$$

$$\sum F_y = N - w = m\bar{a}$$

Reprinted with permission from Overview Case Study Physics, 1991, American Association of Physics Teachers.

**BOX 5-5**
**Visualization/Representation in Chemistry**

Johnstone's 1991 paper entitled "Why Is Science Difficult to Learn?" challenged teachers of chemistry to reexamine their teaching. Chemistry is often taught as a foreign language of sorts, with numbers, letters, and grammatical rules for combining them. Yet chemistry is much more than just balanced equations and molecular formulas. Consider the equation $2Na(s) + Cl_2(g) \rightarrow 2NaCl(s)$. For a chemist, this equation represents the combination of a silvery metal that catches fire in water and a poisonous green gas to create white crystals eaten on french fries. It is the epitome of what excites chemists: the transformation of properties when atoms and molecules form an ordered lattice in which each Na cation is surrounded by six chlorides and vice versa. This equation also encompasses the corners of what is now known as "Johnstone's Triangle:" symbolic, macroscopic, and particulate domains, which are essential to chemistry knowledge (see Chapter 4).

Students, however, typically restrict their efforts to memorizing the symbols for the elements Na and $Cl_2$, balancing the charges of $Na^+$ and $Cl^-$, and balancing the equation. Johnstone noted that to teach chemistry is to help students explore the dyads of the triangle: the relationship between symbols and their corresponding features in particles of molecules and atoms, the connections between the particles and the properties that can be seen and smelled with the human senses, and the language of symbols used to represent elements and compounds.

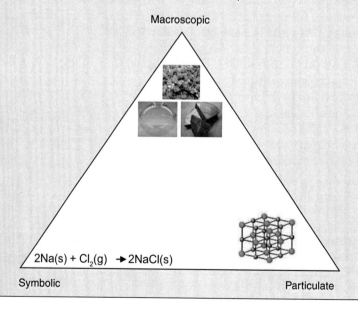

Macroscopic

Symbolic

$2Na(s) + Cl_2(g) \rightarrow 2NaCl(s)$

Particulate

## Biology

Similar to chemistry, some studies suggest that biology students have difficulty translating between alternative abstract diagrammatic representations of the same set of relationships (Novick and Catley, 2007; Novick, Catley, and Funk, 2010; Novick, Stull, and Catley, in press; see Box 5-6). In these studies, students' understanding of the structure of hierarchical diagrams in evolutionary biology appears to be affected by knowledge of biology, by a general perceptual principle from Gestalt psychology called the principle of good continuation, and by a bias to process diagrams in the highly practiced left-to-right reading order.

The principle of good continuation states that a continuous line is perceived as a single entity. Good continuation has been shown to interfere with college students' understanding of the hierarchical structure of cladograms (diagrammatic depictions of descent), and with their ability to reason about the evolutionary relationships depicted when the cladograms were drawn in a way that is commonly found in undergraduate biology textbooks (Catley and Novick, 2008; Novick and Catley, 2007, in press). In follow-up work, a manipulation that broke good continuation at exactly the points in the diagram that mark a new hierarchical level was associated with greatly improved performance (Novick, Catley, and Funk, 2010).

## The Geosciences

Students' use of representations is an emerging area of inquiry in the geosciences. Because relatively few studies exist and most were conducted on a relatively small scale (i.e., in the context of a single course), the strength of the conclusions that can be drawn from geosciences education research on the use of representations is limited. However, taken together, this research lends support to the finding from physics, chemistry, and biology that students have difficulty with specialized representations, and do not reliably interpret them as the maker of the representation had intended (Ishikawa et al., 2005; Swenson and Kastens, 2011). These findings resemble the body of research on students' misconceptions, except that the misunderstandings pertain to a representation rather than a concept. Also similar to the findings from other fields, students have difficulty inferring comparable structures across different geologic representations, such as imagining the internal structure of a block of the earth's crust when given a depiction of external surfaces (Kali and Orion, 1996; Orion et al., 1997).

Because maps are a primary form of representation in the geosciences, methods used in developmental psychology studies of children's understanding of maps (Liben, 1997) provide a potentially useful framework for geoscience education research on the understanding of representations.

**BOX 5-6**
**Visualization/Representation in Biology**

Cladograms, or diagrams that depict evolutionary relationships among a set of taxa, are typically depicted in either a tree or ladder format. Trees are much more common in the evolutionary biology literature, while ladders are slightly more common in high school and college biology textbooks (Catley and Novick, 2008; Novick and Catley, 2007). Novick and Catley (2007) asked college students who had taken at least the first semester of the introductory biology class for majors to translate hierarchical relationships from the nested circles format to the tree and ladder formats and from the tree format to the ladder format and vice versa. The figure in this box shows an example cladogram of each type (the experiment used Latin names of real taxa rather than letters). The figure also shows that students' translation accuracy was reduced whenever the ladder format was involved. Common errors when translating between the tree and ladder formats are consistent with students' interpretation of the long, slanted "backbone" line of ladders as a single entity (i.e., hierarchical level), consistent with the Gestalt principle of good continuation (Kellman, 2000).

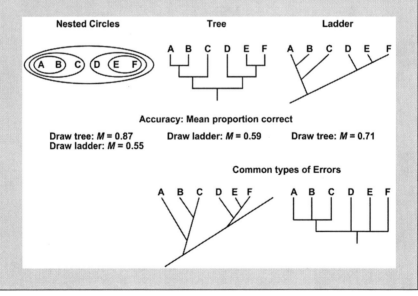

**Nested Circles**          **Tree**          **Ladder**

**Accuracy: Mean proportion correct**

Draw tree: *M* = 0.87          Draw ladder: *M* = 0.59          Draw tree: *M* = 0.71
Draw ladder: *M* = 0.55

**Common types of Errors**

Some geoscience education research has focused on *production tasks*, which are posed in the context of the real world and require students to make or modify a representation. Studies in which participants produce a geological map by direct observation of rocks and structures in nature exemplify this category (Petcovic, Libarkin, and Baker, 2009; Riggs, Balliet, and Lieder, 2009; Riggs, Lieder, and Balliet, 2009). Those studies have revealed variation in the levels of sophistication of advanced geology students' maps, and suggest that more effective traverses of the territory lead to better representations.

*Representational correspondence tasks* provide information in one representation and require the student to transfer or translate that information to another form of representation. A pioneering example is Kali and Orion's (1996) study in which students were shown a block diagram of a three-dimensional geological structure and asked to sketch a vertical profile down through the center of the block. Most of Titus and Horsman's (2009) tasks also fall in this category (see Box 5-7 for a more detailed discussion of this research).

A very limited amount of geoscience education research has been conducted on *metarepresentational tasks*, which require students to explain how a representation works. The work of Liben, Kastens, and Christensen (2011) in which students explain dip and strike (i.e. how to measure the orientation of a sloping planar surface) falls in this category (see "The Role of Spatial Ability in Visualization and Mental Model Formation").

No research yet exists on *comprehension tasks,* which are posed in the context of the representation and require students to respond by performing an action in the real world. An example of this task from the geosciences would be to present students with an interpretive model from the published literature, and ask them to find evidence in an outcrop to support or refute this model (Mogk and Goodwin, 2012).

### The Role of Spatial Ability in Visualization and Mental Model Formation

In many science and engineering disciplines, spatial thinking is a vital component of expertise in the discipline. Some evidence suggests that spatial ability is correlated with learning and reasoning in specific science disciplines, although the strength and importance of the relationship may vary across disciplines (see Hegarty, 2011, for a review). Although these studies have limitations, most importantly concerning causation and causal direction, they do provide evidence for the importance of further investigating the role of spatial abilities in understanding science concepts across disciplines. They also raise issues about how spatial thinking should be promoted, a topic that is considered in some depth by Hegarty (2011).

**BOX 5-7**
**Visualization/Representation in the Geosciences**

In a study of visual penetrative ability, high school students were asked to sketch a specific plan view or profile slice through a three-dimensional volume, using information about the outside of the volume in the form of a block diagram (Kali and Orion, 1996; see the figures in this box). In general, students found this task difficult. Careful examination of incorrect responses and follow-up interviews revealed that partially successful students used specific strategies to "penetrate" the volume, such as continuing a pattern vertically or horizontally (such as depicted in panel 5.a of the figure in this box). Other students, however, gave completely "nonpenetrative" answers, in which they merely copied patterns from the outside of the volume (such as in panel 5.b of the figure in this box) (Kali and Orion, 1996).

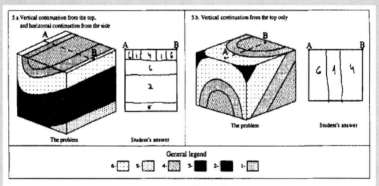

Figure 5. Examples of incorrect answers which indicate an attempt to present interior properties of the structure. Students were required to draw a vertical cross-section through A and B.

Titus and Horsman (2009) extended this line of inquiry to college students, using more spatially challenging structures and tasks, such as drawing structural contours. Titus and Horsman (2009) and Murphy et al. (2011) show that college students' performance on tasks that require visual penetrative ability can be improved by instruction and practice, and that such improvements can be accomplished within the time constraints of standard laboratory instruction.

## Physics

The role of spatial ability in problem solving in mechanics has been an active area of research in physics education. These studies typically focus on the mental processes students use when imagining how the components of a mechanical system (e.g., a spring, gear, or pulley system) move and interact when the system is moving (e.g., Clement, 2009; Hegarty, 1992; Hegarty and Sims, 1994; Schwartz and Black, 1996). This task is referred to as mental animation. The findings from this research suggest that students use both mental images and mathematical analysis to understand how the system is moving and answer the question asked. Other studies on relating force and motion events, interpreting graphs of force and acceleration, extrapolating motion, and inferring the motion of machine components using mental animation have found positive correlations between spatial visualization ability and mechanics problem solving (Hegarty and Sims, 1994; Isaak and Just, 1995; Kozhevnikov, Hegarty, and Mayer, 2002; Kozhevnikov and Thornton, 2006).

## Chemistry

In contrast to physics, research on the relationship between spatial ability and success in chemistry is mixed. Some researchers find spatial ability is a predictor of success on spatial organic chemistry tasks such as assigning configurations and understanding the mechanisms of some reactions (Bodner and McMillian, 1986; Pribyl and Bodner, 1987). Others do not find such correlations, and provide evidence that students of lower spatial ability can successfully use heuristics that do not rely on spatial ability (Abraham, Varghese, and Tang, 2010; Stieff, 2011).

## Engineering

Despite the importance of spatial thinking to engineering, a relatively limited amount of engineering education research exists on this topic. This research reveals some differences related to gender: A 1993 study of 535 first-year Michigan Technological University engineering students showed that females were three times more likely than males to fail the Purdue Spatial Visualization Test on Rotations (PSVT:R) (Sorby, 2009). In addition, engineering education research suggests that spatial skills are correlated with the ability to use computer interfaces to perform database manipulations (Sorby, 2009).

## The Geosciences

The relationship between spatial ability and understanding of geoscience concepts are of great interest to geoscience education researchers

(Liben, Kastens, and Christensen, 2011; Orion, Ben-Chaim, and Kali, 1997). Across many fields of the geosciences, the ability to envision a three-dimensional volume from information that is available in one or two dimensions is essential. This task involves mentally extending the available information into the dimension that is unrepresented, using a combination of spatial thinking ability and knowledge of plausible earth processes. Geologists engage in this activity when they envision what lies behind or between rock outcrops; geophysicists when they infer structures from images of seismic reflections; and oceanographers when they interpolate between stations where seawater temperature and salinity were measured. Kali and Orion (1996) called this ability "visual penetrative ability," and developed a test for it in the context of working with geological structures. As described in Box 5-7, this research provides evidence of students' difficulties with tasks involving visual penetrative ability, reveals variation in high-school students' visual penetrative ability, and shows that the visual penetrative ability of college students can be improved with instruction.

Across the disciplines, very little DBER on spatial ability and representations attends to individual differences (e.g., ability levels) or group differences (e.g., race/ethnicity). A notable exception is a study of dip and strike in the geosciences (Liben, Kastens, and Christensen, 2011). In that study, the researchers purposefully selected participants so as to populate six bins of 20 students each: high-, medium, and low-spatial ability, by male and female. Low-spatial ability (as assessed on the water-level task) and being female were associated with worse performance on the strike and dip tasks (Liben, Kastens, and Christensen, 2011).

*Astronomy*

Considering that astronomy requires learners to imagine a three-dimensional dynamic universe of galaxies and orbiting planets by looking up at a flat sky, it would be reasonable to assume that spatial thinking is an active area of inquiry in astronomy education research. However, systematic research on spatial thinking in astronomy is very limited. Studies are under way to examine the relationships between spatial thinking and astronomy knowledge, but their results have not yet been published in peer-reviewed journals.

## Instructional Strategies and Tools to Improve Students' Spatial Ability and Use of Representations

Given the difficulties students have with spatial thinking and the use of representations, improving these skills is an important part of moving students along the path toward greater expertise, and, in turn, an important focus of DBER. Related research in cognitive science has yielded insights into the reasons for students' difficulties, which also have implications for instruction. Here we discuss these bodies of research as they relate to improving the use of diagrammatic displays, enhancing students' spatial ability, and identifying the role of animations in these tasks.

## Improving the Comprehension of Diagrammatic Displays

In addition to being important tools of discipline, diagrammatic displays—e.g., general-purpose abstract diagrams such as hierarchies and matrices, domain-specific diagrams such as free body diagrams in physics or a drawing of the components of a cell in biology, and graphs of data— are important tools in instruction. These displays can enhance reasoning and problem solving (Larkin and Simon, 1987; Lynch, 1990; Winn, 1989), whether in the sciences or in other disciplines. By storing information externally, diagrams free up working memory that can be used for other cognitive processes. Moreover, the spatial organization of information helps viewers to integrate related information (Hegarty, 2011). These displays can also enable viewers to offload more complex cognitive processes onto simpler perceptual processes. For example, a linear relationship between variables is immediately apparent in a graph but must be laboriously computed from a table of numbers. Similarly, when a display is interactive, people can manipulate the display itself instead of conducting their own computations of the values.

These benefits of graphical displays do not mean that their use is easy or transparent, however, as the previous sections clearly illustrate. One reason diagrams do not automatically facilitate reasoning and problem solving is that their successful use requires knowledge of the conventions underlying their construction (Hegarty et al., 1991). College science is taught by experts who are very familiar with the representations used in their field and may not realize how difficult these representations are for students to master. Instructors may need to provide students with more detailed introductions to the various graphical displays used in science. Such instruction may include the following (Hegarty, 2011):

- explicitly highlighting the relationships between alternative displays of the same or related information,
- explaining how different representations at different levels of abstraction are optimized for various tasks, and
- providing opportunities for extensive practice interpreting and producing appropriate types of representations.

Instruction and practice in making representations might include making "first inscriptions," or representations that are made directly from the represented world or from the raw material of nature (Goodwin, 1994, Mogk and Goodwin, 2012).

A second reason diagrams are not always beneficial is because the relationship between perception and cognition affects how students process diagrams to draw inferences about the represented world. For example, circles and lines seem naturally suited to imply physical objects or locations rather than relationships or motion, respectively (Tversky et al., 2000). When diagrammatic representations are constructed in accordance with these conventions, college students are faster and more accurate at drawing appropriate inferences (Hurley and Novick, 2010). In addition, students often import their highly practiced left-to-right processing strategy from reading printed text to nonlinguistic visual tasks (Fuhrman and Boroditsky, 2010; Nachshon, 1985). In situations where left-to-right processing makes it difficult to interpret relationships in a scientific diagram, simply reflecting the diagram 180 degrees above the vertical axis can improve comprehension (Novick et al., in press). These results suggest that when instructors construct diagrammatic representations to illustrate scientific concepts or principles, they should attend to how people naturally interpret the components from which the representations are constructed.

The Gestalt psychologists identified many other perceptual features that have psychological importance, including two that have particular implications for DBER. The first is good continuation, which states that a continuous line is interpreted as a single entity (discussed in the biology education research section on the role of visualization and spatial ability). The second is spatial proximity, which states that items that are closer together tend to be processed together. This principle has been shown to play an important role in students' understanding of cladograms (Novick and Catley, in press) and mathematical equations (Landy and Goldstone, 2007). Whether spatial proximity facilitates or hinders students' performance depends on whether the conclusion based on proximity supports or conflicts with the correct response. In some situations, visual representations may be constructed in such a way that spatial proximity is a useful clue rather than a conflicting signal, as when a flow chart or network diagram is constructed with functionally related subsystem elements clustered together within a larger depicted system. In other cases, however, instructors probably need to tackle the issue directly, such as through the use of refutational instructional strategies (Hynd et al., 1994; Kowalski and Taylor, 2009), explaining why spatial proximity is not relevant and what structural features of the diagram instead provide the critical information needed for responding (Novick, Catley, and Schreiber, 2010).

## Improving Spatial Ability

Research from engineering and the geosciences suggests that spatial ability can be improved through instruction. One review article (Sorby, 2009) summarizes more than a decade of applied research—including several longitudinal studies—aimed at identifying practical methods for improving three-dimensional spatial skills, especially for female engineering students. That article identifies strategies that appear to be effective in developing spatial skills. In the longitudinal studies, first-year engineering majors of low spatial ability (as measured by the PSVT:R) who took a specially designed, multimedia training course improved their spatial skills, earned higher grades, and persisted in the university at greater rates than students of similar spatial ability who did not take the course. The multimedia software and workbook that were part of the course materials were shown to improve students' spatial skills. In addition, sketching was consistently identified as an important component of spatial skill development (Sorby, 2009).

Some research indicates that geoscience education involving visually rich materials can improve students' spatial visualization skills on domain-specific tasks and general spatial tasks in ways that are practical to incorporate into instructional practices (Ozdemir et al., 2004; Piburn et al., 2005; Sawada et al., 2002; Titus and Horsman, 2009) (see Box 5-7 for an example). In some of those studies, including a quasi-experimental study (Piburn et al., 2005), an initial gap between males and females in spatial visualization ability (as measured by the surface development task) diminished after visualization-rich instruction in physical geology (Piburn et al., 2005; Sawada et al., 2002).

### The Influence of Animated and Static Visualizations on Conceptual Understanding

In addition to using specialized representations, expert scientists and engineers use technologies such as animations, interactive computer visualizations and virtual models to aid their work. These technologies have enormous potential for promoting representational competence and spatial thinking in science (Hegarty, 2011). Indeed, several quasi-experimental studies comparing the efficacy of static visuals with two- and three-dimensional animations of molecular structures and processes appear to improve student learning of stereochemistry, which concerns the spatial arrangement of atoms within molecules (Abraham, Varghese, and Tang, 2010; Aldahmash and Abraham, 2009; Sanger and Badger, 2001). These studies suggest that three-dimensional animations improve learning more than two-dimensional animations, which, in turn, are better than static representations. The use of animations in biology classes also has been shown to increase retention of content knowledge in the short and long-term (Harris et al., 2009; McLean et al., 2005; O'Day, 2007). The animations in those studies included a three-dimensional animation of protein synthesis (McLean et al., 2005), a narrated animation related to a complex signal transduction pathway (O'Day, 2007), and a combination of a molecular imaging program and handheld physical models related to molecular structure and function (Harris et al., 2009). Although this research shows that students can learn more from animations than from static images, McLean et al. (2005) caution that "learning is best achieved when an animation is coupled with a lecture, because this combination provides a reference from which students can appreciate the knowledge presented in the animation" (p. 170).

The recommendation of McLean et al. (2005) might reflect the fact that visual technologies are typically more complex than static visualizations. Thus, these technologies may require greater spatial ability and other aspects of representational competence for their successful use. It is not a foregone conclusion, therefore, that all animations will necessarily be beneficial for learning. In this regard, some cognitive science research comparing static and animated displays of dynamic biological and mechanical processes has found no benefit in performance when animations were used (Tversky, Morrison, and Betrancourt, 2002).

Sometimes, animations present information too quickly for students to accurately perceive and comprehend. Although the pace of animations can easily be slowed, other problems are not so readily fixed. For example, animations of a mechanical system can give students the illusion that they understand the information being presented and may even cause them to "see" what they believe is true rather than what is actually presented (Kriz and Hegarty, 2007). Some research has found that animations showing how a mechanical system works were more effective in promoting learning when students had to predict how they thought the machine works before viewing the animation (Hegarty, Kriz, and Cate, 2003). The authors hypothesize that the prediction task improves the effectiveness of the animation by making students aware of what they do not understand, thus cueing them to the type of information to extract from the animation. This finding is consistent with those discussed in Chapter 6 about the benefits of interactive lecture demonstrations. One conclusion to draw is that although animated, three-dimensional, and interactive visualizations might seem to rely less on visualization skills, research to date has suggested that they actually require visualization skills for their use (Hegarty, 2011).

## Summary of Key Findings on Spatial Thinking and the Use of Representations

- *How students create, use, interpret, and translate between graphical and mathematical representations provides insight into their understanding of important concepts in a discipline. DBER highlights discipline-specific challenges that students face when using such representations.*

- *Although equations, graphical displays, and other representations may seem easy to understand for undergraduate faculty who are domain experts, college students have difficulty extracting information from these representations, and constructing appropriate representations from existing information. College students also have difficulty relating and translating among different representations of the same entity or phenomenon.*

- *There is contradictory evidence about the relationship between spatial ability and performance in science. Consistent with findings from cognitive science that students with low-spatial ability especially have difficulty relating two- and three-dimensional representations, some DBER studies show a relationship between measures of spatial ability and success on specific science or engineering tasks. Other studies do not provide evidence of that relationship.*

- *The evidence on the effectiveness of animations is mixed: The use of animations has been shown to enhance learning in some circumstances, and to be ineffective or even detrimental to student learning in other situations.*

## Directions for Future Research on Spatial Thinking and the Use of Representations

DBER and cognitive science have yielded many useful insights into how students use mathematical and graphical representations, but important gaps remain. For example, the research community, instructors, and those who develop representations would benefit from a deeper understanding of students' use of representations as tools to enhance their learning, and studies along these lines should leverage what is already known about the basic cognitive and perceptual processes that students use to comprehend graphical representations.

The role of spatial ability also needs clarification. Spatial ability may be measured in many different ways, any one of which may be more or less relevant to any specific science or engineering task. Although several authors have proposed that many tasks (e.g., rotation tasks of three-dimensional models) require mental imagistic models, others have shown that many students use heuristics and other strategies that do not employ visualization skills and are able to move flexibly between such strategies as needed. In addition to clarifying the overall role of spatial ability, it would be useful to evaluate the contributions of large-scale and small-scale spatial ability to learning in physics, chemistry, engineering, biology, the geosciences, and astronomy. DBER has not yet examined these different spatial abilities.

The research base on promoting students' understanding of and facility with domain-specific representations is less robust. DBER does not provide conclusive evidence about how instructors, illustrators, and authors should design representations for maximum effect, or what the optimal representations are for a given situation. Moreover, additional research is needed to identify the range of instructional approaches that help students use mathematical and graphical representations to enhance their knowledge and understanding. For example, does designing and constructing representations affect students' understanding differently than merely interpreting existing representations, and if so, how? Given the increasing use of technology, more research is needed on the educational efficacy of computer animations, simulations, and other technology-enhanced techniques that aid with visualization and representations, and the conditions under which those techniques are effective.

Representations vary within and across disciplines. As one example, the nature of the representations used in geoscience education varies enormously on multiple important dimensions, including the use of spatial representations to represent nonspatial data (Dutrow, 2007; Kastens, 2009, 2010; Libarkin and Brick, 2002). This variation presents a challenge to developing a research agenda for the use of visualizations and representations in undergraduate science and engineering education, because research using any specific representation may not be generalizable to other representations.

# 6

# Instructional Strategies

In addition to the strategies described in Chapters 4 and 5 to promote conceptual change and improve students' problem solving and use of representations, scientists and engineers want to provide the most effective overall learning experiences to help students acquire greater expertise in their disciplines. To some extent, those experiences are constrained by institutional context. Undergraduate lecture halls and laboratories provide much of the infrastructure for teaching students in science and engineering. One compelling question is how best to use those resources. An undergraduate course may be structured around traditional lectures offered two or three times weekly along with a laboratory experience. Some scientists and engineers want to explore alternatives to this traditional format. If they were to depart from the lecture-plus laboratory format, then according to discipline-based education research (DBER), which teaching options are most promising? More importantly, which options are backed by evidence for their effectiveness in fostering student learning?

A significant portion of DBER focuses on measuring the impact of instructional strategies on student learning and understanding. In this chapter, we summarize that research, discussing the three most common settings for undergraduate instruction—the classroom, the laboratory, and the field—and the effects of instructional strategies on different student groups.

## OVERVIEW OF DISCIPLINE-BASED EDUCATION RESEARCH ON INSTRUCTION

As stated in Chapter 1, two long-term goals of DBER are to help identify and measure appropriate learning objectives and instructional approaches that advance students toward those objectives, and identify approaches to make science and engineering education broad and inclusive. This research is motivated, in part, by ongoing concerns that undergraduate science and engineering courses are not providing students with high-quality learning experiences or attracting students into science and engineering degrees (President's Council of Advisors on Science and Technology, 2012). Indeed, a seminal three-year, multicampus survey examined the reasons undergraduate students switch from science, mathematics, and engineering majors to nonscience majors (Seymour and Hewitt, 1997). The survey revealed that nearly 50 percent of undergraduates who began in science and engineering shifted to other majors. Their reasons for doing so were complex and numerous, but pedagogy ranked high among their concerns. In fact, poor faculty pedagogy was identified as a concern for 83 percent of all science, mathematics, and engineering students. Forty-two percent of white students cited poor pedagogy as the primary factor in their decision to shift majors, compared with 21 percent of non-Asian students of color, who tended to blame themselves and suffered a substantial loss of confidence in leaving the sciences (Seymour and Hewitt, 1997).

Recognizing these challenges, many institutions are working to identify effective approaches to improve undergraduate science and engineering education (Association of American Universities, 2011). DBER, by systematically investigating learning and teaching in science and engineering and providing a robust evidence base for new practices, is playing a critical role in these efforts.

### Research Focus

Most DBER studies on instructional strategies are predicated on the assumption that students must build their own understanding in a discipline by applying its methods and principles, either individually or in groups (Piaget, 1978; Vygotsky, 1978). Consequently, with some variations, these studies typically examine student-centered approaches to learning, often comparing the extent to which student-centered classes are more effective than traditional lectures in promoting students' understanding of course content.

A student-centered instructional approach places less emphasis on transmitting factual information from the instructor, and is consistent with the shift in models of learning from information acquisition (mid-1900s) to knowledge construction (late 1900s) (Mayer, 2010). This approach includes

- more time spent engaging students in active learning during class;
- frequent formative assessment to provide feedback to students and the instructor on students' levels of conceptual understanding; and
- in some cases, attention to students' metacognitive strategies as they strive to master the course material.

The extent to which DBER on instructional practices is explicitly grounded in broader research on how students learn varies widely. The committee's analysis revealed that either implicitly or explicitly, the principle of active learning has had the greatest influence on DBER scholars and their studies. With a deep history in cognitive and educational psychology, this principle specifies that meaningful learning requires students to select, organize, and integrate information, either independently or in groups (Jacoby, 1978; Mayer, 2011; National Research Council, 1999). In addition, the framework of cognitive apprenticeship drives many instructional reforms in physics and thus can help to explain research findings about the success of those reforms. As described in Chapter 5, cognitive apprenticeship is based on the idea that complex skills depend on an interlocking set of experiences and instruction whose efficacy, in turn, depend on the learner and the community of practitioners with whom the learner interacts (Brown, Collins, and Duguid, 1989; Yerushalmi et al., 2007).

Although some DBER is guided by learning theories and principles, reports of DBER studies are typically organized around instructional setting. Following that convention, we organize our synthesis of DBER on instruction by setting—classroom, laboratory, and field—before considering the effects of instructional strategies on different groups.

## Methods

Most of the available research on instruction is conducted in introductory courses. Sample sizes range from tens of students to several hundred students. The preponderance of this research is conducted in the context of a single course or laboratory—often by the instructor of that course, and sometimes comparing outcomes across multiple sections of that course. Fewer studies are conducted across multiple courses or multiple institutions.

Many studies use pre- and post-tests of student knowledge (often with a comparison or control group) to assess some measure of learning gains for one course, typically lasting one semester. These gains often are measured with concept inventories developed for aspects of the discipline or other specialized assessments (see Chapter 4 for a discussion of concept inventories), or with course assignments or exams. Fewer studies measure longer-term gains, or other outcomes such as student attitudes and motivation to study the discipline.

## INSTRUCTION IN THE CLASSROOM SETTING

Understandably, most DBER on instructional strategies centers on the classroom setting. The reviews of DBER commissioned for this study (Bailey, 2011; Dirks, 2011; Docktor and Mestre, 2011; Piburn, Kraft, and Pacheco, 2011; Svinicki, 2011; Towns and Kraft, 2011), along with other syntheses (e.g., Allen and Tanner, 2009; Hake, 1998; Handelsman, Miller, and Pfund, 2007; Prince, 2004; Ruiz-Primo et al. 2011; Smith et al., 2005; Wood, 2009) consistently support the view that adopting various student-centered approaches to classroom instruction at the undergraduate level can improve students' learning relative to lectures that do not include student participation. A limited amount of research suggests that even incremental changes toward more student-centered approaches can enhance students' learning (Derting and Ebert-May, 2010; Knight and Wood, 2005).

Research from the different fields of DBER reveals some nuances and variations on this theme, which we explore in this section. We have organized this discussion by instructional strategy rather than by discipline because these strategies in themselves are not discipline-specific, and most are implemented in similar learning environments. We include discipline-specific discussions under each strategy where that research was available.

### Making Lectures More Interactive

Most undergraduate science and engineering classes are taught in a lecture format. Although traditional lectures can be effective for some students (Schwartz and Bransford, 1998), instructors have a variety of options at their disposal to make lectures more interactive and enhance their effectiveness. These options range in scope and complexity from slight modifications of instructional practice—such as beginning a lecture with a challenging question for students to keep in mind—to devoting most of the instructional time to collaborative problem solving. Research on making lectures more interactive is a significant focus of DBER. Overall, the committee has characterized the strength of the evidence on making lectures more interactive as strong because of the high degree to which the findings converge, albeit from many studies that were conducted in the context of a single course using a wide variety of measurement tools. This section discusses several options for making lectures and small discussion groups more interactive. Most of these approaches involve enhancing or refining—rather than completely eliminating—the lecture format.

#### Encouraging Student Participation

Interactive lectures involve students in learning the material, often requiring them to think and apply the content that is covered during class. Several geoscience education research studies have examined the effectiveness of interactive lectures. One study (Clary and Wandersee, 2007) tested a model of integrated, thematic instruction in the introductory geology lecture. Students in the experimental condition did an in-lecture "mini-lab" with petrified wood and discussed their observations in on-line discussion groups. Pre-test/ post-test application of a researcher-developed survey showed statistically greater gains in the experimental group than in two control groups. Other research examining the use of ConcepTests (short, formative assessments of a single concept), Venn diagrams constructed with student input, and analysis of geologic images during lecture has shown significant differences between control and experimental groups; students who experienced the interactive strategies earned higher exam scores (McConnell, Steer, and Owens, 2003).

Interactive lecture demonstrations are another strategy for encouraging student participation. With this approach, students (1) make predictions about the outcome of a physical demonstration that the instructor conducts in class, (2) explain this prediction with peers and then with the class, (3) observe the event, and (4) compare their observations to their predictions (Sokoloff and Thornton, 2004). Some research on interactive lecture demonstrations indicates that they can improve students' understanding of foundational physics concepts as measured by the Force and Motion Conceptual Evaluation (Sokoloff and Thornton, 1997). Other research suggests that the prediction phase (consistent with conceptual-change models) is particularly important to the success of an interactive lecture demonstration (Crouch et al., 2004). Similarly, chemistry education research shows that students who were allowed to work in small groups to make predictions about lecture demonstrations showed significant improvements on tests over students who merely observed demonstrations (Bowen and Phelps, 1997).

Another approach is to adapt lectures based on student responses to pre-class or in-class work. The most familiar pre-lecture method is Just-in-Time Teaching. With this approach, students read and answer questions or solve homework problems before class and submit their work to the instructor electronically, with enough time for the instructor to modify the lecture to target student weaknesses or accommodate their interests (Novak, 1999). A moderate amount of evidence suggests that Just-in-Time Teaching is effective in teaching some physics concepts, such as Newton's Third Law (Formica, Easley, and Spraker, 2010), and is associated with positive attitudes about introductory geology (Linneman and Plake, 2006; Luo, 2008). In biology, Just-in-Time Teaching has been associated with improved student preparation for classes and more effective study habits; students also preferred this format to traditional lectures (Marrs and Novak, 2004).

Other versions of pre-lecture assignments have been associated with gains in student learning. As one example, Multimedia Learning Modules have been associated with improved course performance in physics (Stelzer et al., 2009). In a large introductory biology course for majors, students who participated in Learn Before Lecture (a simpler approach than Just-in-Time Teaching) performed significantly better than students in traditional courses on Learn Before Lecture-related exam questions, but not on other questions (Moravec et al., 2010).

Although arguably less common, approaches that involve real-time adjustment of instruction also appear to have the potential to improve student learning and performance. In a quasi-experimental study in the geosciences, students in interactive courses were given brief introductory lectures followed by formative assessments that triggered immediate feedback and adjustment of instruction. These students showed a substantial improvement in Geoscience Concept Inventory scores (McConnell et al., 2006).

Audience response systems ("clickers") are a different approach to encouraging greater student participation in large-enrollment courses. Clickers are small handheld devices that allow students to send information (typically their response to a multiple choice question provided by the instructor) to a receiver, which tabulates the classroom results and displays the information to the instructor. The value of clickers for in-class formative assessment has been debated. Some biology instructors have reported high student approval and enhanced learning using clickers (e.g., Smith et al., 2009; Wood, 2004), while others have found them less useful and have discontinued their use (Caldwell, 2007). Research in chemistry and astronomy suggests that learning gains are only associated with applications of clickers that incorporate socially mediated learning techniques, such as those discussed in the next section (Len, 2007; MacArthur and Jones, 2008). Overall, the research on clickers indicates that technology itself does not improve outcomes, but how the technology is used matters more (e.g., Caldwell, 2007; Keller et al., 2007; Lasry, 2008).

Regarding clickers—as regarding instruction more broadly—DBER has not yet systematically used learning theory principles to examine whether certain strategies are more effective for different populations of students, or analyzed the conditions under which those strategies are successfully implemented. However, several authors have offered suggestions for best practices with clicker technology (Beatty et al., 2006; Caldwell, 2007; Smith et al., 2009; Wieman et al., 2008), including posing formative assessment questions at higher cognitive levels and socially mediated conditions for learning such as allowing students to discuss their responses in groups before the correct answer is revealed.

### Involving Students in Collaborative Activities

Many transformed courses (i.e., courses in which instructors are using student-centered approaches) incorporate in-class activities where students collaborate with each other. Consistent with research from science education and educational psychology, DBER has shown that these activities enhance the effectiveness of student-centered learning over traditional instruction (e.g., Armstrong, Chang, and Brickman, 2007; Johnson, Johnson, and Smith, 1998; Smith et al., 2009, 2011; Springer, Stanne, and Donovan, 1999). Moreover, collaborative learning has been shown to improve student retention of content knowledge (Cortright et al., 2003; Rao, Collins, and DiCarlo, 2002; Wright and Boggs, 2002). However, it is important to remember that collaborative learning is not inherently effective, and this approach can be implemented ineffectively (Slavin, Hurley, and Chamberlain, 2003). In this vein, DBER does not yet provide conclusive evidence about the conditions under which these strategies are effective, and for which students.

Think-Pair-Share is a straightforward form of in-class collaborative activity—widely used in K-12 education—that is also referred to as informal cooperative learning (Johnson, Johnson, and Smith, 2011; Smith, 2000). With this approach, the instructor poses a question, often one that has many possible answers; asks students to formulate answers, share their answers, and discuss the question with their group; elicits answers again; and engages in a class-wide discussion. The use of informal groups in this way has been associated with improvements in a variety of outcomes, including achievement, critical thinking and higher-level reasoning, students' understanding of others' perspectives, and attitudes about their fellow students, instructors, and the subject matter at hand (Johnson, Johnson, and Smith, 2007, 1998; Smith et al., 2005). Instructors adapt Think-Pair-Share in various ways. Some geoscience education researchers have followed brief introductory lectures with interactive sessions during which students discussed ideas in groups and completed worksheets based on the misconceptions literature. On average, students who participated in the interactive sessions scored higher on tests than students who received only lecture, even when taught by the same instructor during the same semester (Kortz, Smay, and Murray, 2008).

In chemistry, a number of initiatives that stress socially mediated learning have been widely adopted and adapted. In POGIL (Process-Oriented Guided Inquiry Learning),[1] students work together in small groups on guided inquiry activities to learning content and science practices. PLTL (Peer Led Team Learning)[2] uses peer-team leaders in out-of-class team problem-solving sessions. Both POGIL and PLTL have developed large communities of practice, and there is some evidence that they can improve student outcomes. One mixed-methods study reported significantly improved

---

[1]For more information, see http://www.pogil.org [accessed April 13, 2012].
[2]For more information, see http://www.pltl.org [accessed April 13, 2012].

outcomes for organic chemistry students in PLTL sections on all course exams and finals, compared with students who learned through traditional lecture courses (Tien, Roth, and Kampmeier, 2002). Other studies have shown that a combination of PLTL and POGIL improved test scores for a cohort of students in general chemistry (Lewis and Lewis, 2005). However, much more research remains to be done to investigate how these pedagogies can best be implemented, how different student populations are affected, and how the fidelity of implementation—that is, the extent to which the experience as implemented follows the intended design—affects outcomes.

To explore the common view that group learning is pragmatically impossible in large-enrollment courses, some astronomy education researchers created and systematically studied a series of collaborative group activities modified specifically for large-enrollment courses known as ASTRO 101. We have characterized the strength of this evidence as limited because relatively few studies exist and the results have not been independently replicated. Studies of these activities reveal that students can learn more when collaborative group activities are added to traditional lecture and that they enjoy the collaborative learning experience more than traditional courses (Adams and Slater, 1998, 2002; Skala, Slater, and Adams, 2000). In addition, female-only learning groups performed better than heterogeneous groups in these activities (Adams et al., 2002). Survey responses, course evaluations, and exam performance in large-enrollment (600 students) oceanography courses have also revealed an increased interest in science as well as improvements in subject-matter learning, information recall, analytical skills, and quantitative reasoning for students who were taught with cooperative learning and collaborative assessments (Yuretich et al., 2001).

In addition to being used in large lectures, collaborative activities also are used to make smaller discussion sections more interactive. In physics, Cooperative Group Problem Solving requires students to work in formal, structured groups on specifically designed tasks called context-rich problems (Heller and Heller, 2000; Heller and Hollabaugh, 1992). The design of this highly structured approach is based on research on cooperative learning, a popular method in K-12 education (Johnson, Johnson, and Holubec, 1990; Johnson, Johnson, and Smith, 1991). A limited amount of evidence at the undergraduate level suggests that this approach can contribute to improved conceptual understanding and problem-solving skills (Heller and Hollabaugh, 1992; Heller, Keith, and Anderson, 1992; Hollabaugh, 1995) (see Box 6-1 for a description of other collaborative models used in physics in which a key feature is changing the learning space). Findings from a study in chemistry also indicated that cooperative group problem solving improved students' problem-solving abilities by about 10 percent, and that this improvement was retained when students returned to individual problem-solving activities (Cooper et al., 2008). In that study, the only students who did not benefit

**BOX 6-1
Changing the Learning Space:
Some Examples from Physics**

Several physics education reforms have involved redesigning the learning space. Based on the model of cognitive apprenticeship (see Chapter 5), these redesigns also involve dramatic changes to the way physics is taught, reducing the amount of lecturing and often integrating laboratory and lecture. Some examples include the following:

*Workshop Physics.* Developed at Dickinson College, *Workshop Physics* taught university physics entirely within the laboratory, using the latest computer technology. Students preferred workshop courses, and students in these courses generally outperformed students in traditional courses on conceptual exams but not in problem solving (Laws, 1991, 2004).

*Studio Physics.* Developed at Rensselaer, *Studio Physics* redesigned teaching spaces to accommodate an integrated lecture/laboratory course. Early studies showed little improvement in students' conceptual understanding or problem-solving skills, despite the popularity of the innovation. Later implementations, which added research-based curricula, resulted in improved learning of content over traditional courses (Cummings et al., 1999; Sorensen et al., 2006) but not always improvements in problem solving (Hoellwarth, Moelter, and Knight, 2005).

*SCALE-UP.* Developed at North Carolina State University, the *Student-Centered Active Learning Environment for Undergraduate Programs (SCALE-UP)* begins with a redesign of the classroom. Each room holds approximately 100 students, with round tables that accommodate 3 laptops and 9 students, whiteboards on several walls, and multiple computer projectors and screens so every student has a view. Students engage in hands-on activities and with computer simulations, work collaboratively on problems, and conduct hypothesis-driven experiments. SCALE-UP students have better scores on problem-solving exams and concept tests, slightly better attitudes about science, and less attrition than students in traditional courses (Beichner et al., 2007; Gaffney et al., 2008).

from this activity were students with the lowest scores on a logical thinking test who were paired with students of similar ability.

Teasing apart the benefits of collaborative group versus individual problem-solving practice is difficult, as is following changes in problem-solving ability over time, particularly in large classes. Some recent work has been done on the development and validation of tools for comparing collaborative and individual problem-solving strategies in large (60-100 students) biochemistry courses, with students discussing ill-defined problems in small online groups (Anderson, Mitchell, and Osgood, 2008), and then working through individual electronic exams based on similar, but not identical, problems (Anderson et al., 2011).

## Other Instructional Strategies

Some DBER exists on other popular instructional strategies that are not necessarily interactive. We have characterized the strength of conclusions that can be drawn from this evidence as limited because relatively few studies exist and the findings across disciplines are contradictory. For example, in traditional and student-centered classes alike, analogies and explanatory models are widely used pedagogical tools to help students see similarities between what they already know and unfamiliar, often abstract concepts (Clement, 2008). Some physics education research suggests that use of analogies during instruction of electromagnetic waves helped students generate inferences, and that students taught with the help of analogies outperformed students who were taught traditionally (Podolefsky and Finkelstein, 2006, 2007a). Further research indicates that blending multiple analogies to convey wave concepts can lead to better student reasoning than using single analogies or standard abstract representations (Podolefsky and Finkelstein, 2007b). A possible explanation for this finding is that using multiple analogies may have helped learners to see the general pattern across the separate analogies (Gentner and Colhoun, 2010), rather than becoming overly attached to the specific features of any one analogy. This result echoes findings from cognitive science that multiple analogies facilitate problem solving because they help solvers to construct a general schema for the common underlying solution procedure (Catrambone and Holyoak, 1989; Gick and Holyoak, 1983; Novick and Holyoak, 1991; Ross and Kennedy, 1990).

In contrast to findings from physics education research, a series of chemistry education research studies identifies the challenges of using analogies for college students who had successfully completed at least one biochemistry course (Orgill and Bodner, 2004, 2006, 2007). In those studies, faculty used analogies to identify similar features between the already-known concept and the concept to be learned, with the goal of facilitating the transfer of knowledge from one setting to another. However, the instructors often did not identify where the analogy broke down or failed to be useful. As a result, students overgeneralized the features of the known situation, thinking that all features were represented in the target. This overgeneralization impaired student learning.

Another approach in teaching science and engineering is to present abstract concepts and then follow them with a specific worked example (sometimes called a "touchstone example") to illustrate how the concepts are applied to solve problems. With this approach, students' understanding of the concept often becomes conflated with the particulars of the example that is used. As a result, students may have difficulty separating the solution from the specifics of a particular problem, which may limit their ability to apply knowledge of the concept in other settings. This phenomenon is known as the "specificity effect" and has been demonstrated in several physics education research studies (Mestre et al., 2009) as well as basic studies in cognitive science.

## Supplementing Instruction with Tutorials

The tutorial approach is a common instructional innovation in physics and astronomy, and represents a significant area of research and development for physics and astronomy education research. With a tutorial approach, instructors are provided with a classroom-ready tool to target a specific concept, elicit and confront tenacious student misconceptions, create learning opportunities, and provide formative feedback to students.

The University of Washington physics education research group has developed several *Tutorials in Introductory Physics* (McDermott and Shaffer, 2002), and numerous studies have demonstrated that these tutorials significantly improve student understanding of the targeted concepts and of scientific reasoning more generally (see review by Docktor and Mestre, 2011, for a detailed listing of relevant publications). The success of the University of Washington tutorials has inspired other research groups to create and evaluate tutorial-style learning interventions (e.g., Elby, 2001; Steinberg, Wittmann, and Redish, 1997; Wittmann, Steinberg, and Redish, 2004, 2005). In physics, these adaptations are predominantly used in a recitation or discussion section.

Astronomy education researchers have successfully modified the tutorial approach to be used in a lecture classroom environment. For example, *Lecture-Tutorials for Introductory Astronomy* (Prather et al., 2004, 2007) is a widely used series of short-duration, highly focused, highly structured learning activities. Instructors lead students through a purposeful sequence of carefully constructed questions designed to move the learner toward a more expert-like understanding. Several studies have shown that the lecture-tutorial approach is more effective than lecture-dominated courses in improving students' understanding in astronomy (Alexander, 2005; Bailey and Nagamine, 2009; Lopresto, 2010; Lopresto and Murrell, 2009). One study of multiple introductory science courses across multiple institutions revealed that adaptations of the astronomy approach for introductory geoscience courses improved students' test scores in those courses (Kortz, Smay, and Murray, 2008).

## INSTRUCTION IN THE LABORATORY SETTING

Learning science and engineering takes place not just in classrooms, but also in laboratories[3] and in the field. Well-designed laboratories can help students to develop competence with scientific practices such as experimental design; argumentation; formulation of scientific questions; and use of discipline-specific equipment such as pipettes, microscopes, and volumetric glassware. However, laboratories that are designed primarily to reinforce lecture material do not necessarily deepen undergraduate students' understanding of the concepts covered in lecture (Elliott, Stewart, and Lagowski, 2008; Herrington and Nakhleh, 2003; Hofstein and Lunetta, 1982; Kirschner and Meester, 1988 Lazarowitz and Tamir, 1994; White, 1996). Indeed, a 2004 review of more than 20 years of research on laboratory instruction found "sparse data from carefully designed and conducted studies" to support the widely held belief that laboratory learning is essential for understanding science (Hofstein and Lunetta, 2004, p. 46).

Relatively few DBER studies focus on the laboratory environment. We have characterized the strength of evidence as moderate in physics because the research base includes a combination of smaller-scale studies (e.g., a single course or section) and studies that have been conducted across multiple courses or institutions, with general convergence of findings. In chemistry, engineering, biology, the geosciences, and astronomy, the strength of the conclusions that can be drawn from this research is limited.

### Physics

One of the criticisms of traditional laboratory manuals is that they do not reflect what scientists actually do: develop hypotheses, design and conduct experiments, make decisions about measurement error versus equipment sensitivity, and report their findings. Several reformed physics

---

[3]It was beyond the scope of this committee's charge to define what constitutes a laboratory course (see National Research Council [2006] for a definition of laboratory experiences for K-12 education). Recognizing the wide range of laboratory experiences—and the variations within and across disciplines—in this report, we describe what is commonly practiced in each discipline by using the operational definitions of laboratory employed in the research we reviewed.

curricula include laboratory experiences that are aligned with scientific practices (see, for example, *Investigative Science Learning Environment* [Etkina and Van Heuvelen, 2007], *Physics by Inquiry* [McDermott et al., 1996a, 1996b)], and *Modeling Instruction* [Brewe, 2008]). In these laboratory exercises, students record observations, develop and test explanations, refine existing models, and build and refine their own causal models through experimentation.

Studies of specific curricular innovations show that these types of laboratories are more effective than traditional laboratories for developing students' ability to design experiments, collect and analyze data, and engage in more authentic scientific communication (Etkina et al., 2006, 2010; Karelina and Etkina, 2007). These laboratories also contribute to positive attitudes about introductory physics, as measured by the Colorado Learning Attitudes about Science Survey (Brewe, Kramer, and O'Brien, 2009), in contrast to most other introductory physics courses (Redish, Steinberg, and Saul, 1998). A limited amount of evidence suggests that some of these benefits may extend beyond the laboratory setting. For example, one study showed that the skills learned in a reformed physics laboratory can transfer to novel tasks in biology (Etkina et al., 2010). In another study, students in a reformed laboratory outperformed their peers from traditional laboratories on course exam problems (Thacker et al., 1994).

Some physics education research has examined the use of technology in the laboratory setting. One curriculum, *RealTime Physics Active Learning Laboratories*, targets known misconceptions by using microcomputer-based technologies to instantly analyze formative data and provide immediate feedback to the student. Studies of *RealTime Physics* show gains on the Force Motion Concept Inventory (Sokoloff and Thornton, 1997) over traditional laboratories, although the value of the instantaneous feedback on improving students' learning is debated (Beichner, 1990; Brasell, 1987; Brungardt and Zollman, 1995). A limited amount of evidence also suggests that video-based laboratories, where students either create their own videos of motion in the laboratory or use provided videos such as a space-shuttle launch and then analyze the videos using specific software programs, can improve students' understanding of kinematics and kinematics graphs (Beichner, 1996). In addition, interactive computer simulations of physical phenomena can lead to improved student performance on laboratory reports, exam questions, and performance tasks (e.g., assembling real circuits) over traditional instruction (Finkelstein et al., 2005).

## Chemistry

The chemistry laboratory is where the properties and reactions between chemicals become visible, and where chemists extrapolate the properties of compounds to their molecular structure. For chemistry faculty, the laboratory is integral to learning chemistry. Given the expense of laboratory instruction, however, the question of whether students can learn chemistry without laboratories is asked with increasing frequency by department chairs and faculty administrators.

Despite its importance in the curriculum, the role of the chemistry laboratory in student learning has gone largely unexamined. The research that has been done has investigated faculty goals for laboratory learning, the role of graduate students as teaching assistants in the laboratory, experiments to restructure the laboratory with an inquiry focus, and students' interactions with instrumentation in the laboratory.

An interview study of chemistry faculty revealed that faculty goals vary for connecting laboratory to lecture, promoting students' critical thinking, providing experiences with experimental design, and teaching students about uncertainty in measurement (Bruck, Towns, and Bretz, 2010). Research on students' experiences in general chemistry (Miller et al., 2004) and analytical chemistry (Malina and Nakhleh, 2003) suggests that such variation can influence students' views of laboratory learning. Depending on how faculty members structure the laboratory experiment and assess student learning, students can view instruments simply as objects, without any knowledge of their internal workings, or as useful tools for collecting evidence about the behavior of molecules and their properties.

Domin (1999) has characterized inquiry in chemistry laboratories as ranging from deductive experiences ("explain, then experiment") to inductive experiments ("experiment, then explain"). To explore learning along this continuum, Jalil (2006) designed a laboratory course with both kinds of experiments, finding that although students initially preferred deductive experiments, they eventually came to value the inductive approach because the experiments provided them with knowledge for subsequent learning in lecture. Although the label "inquiry" is often synonymous with inductive experiments, one analysis (Fay et al., 2007) found that neither commercially published laboratory manuals nor peer-reviewed manuscripts that self-identify as "inquiry" score very high on Lederman's rubric of scientific inquiry, which was designed to assess the level of scientific inquiry occurring in high-school science classrooms. This research has been extended to other disciplines with similar results (Whitson et al., 2008).

Regarding the effect of laboratories on learning, emerging evidence suggests that students in an open-ended, problem-based laboratory format improve their problem-solving skills (Sandi-Urena et al., 2011, in press). The science writing heuristic—which combines an instructional technique to improve the flow of activities during an experiment with an alternative format for writing laboratory reports—is another approach to improve student learning. Research has shown that students who were taught by teaching assistants who implemented the science writing heuristic appropriately showed significant improvements on their lecture exam scores (Rudd, Greenbowe, and Hand, 2007). In contrast, traditional laboratories that confirm the knowledge students may already possess do not appear to increase their understanding or retention (Gabel, 1999; Hart et al., 2000; Hofstein and Mamlok-Naaman, 2007).

## Biology

Biology education research studies on instruction in the laboratory setting typically examine the outcomes of inquiry-based laboratories, often in comparison to traditional laboratories. The design of inquiry-based laboratories is based on the concept of the learning cycle, in which students pose questions, confront their misconceptions, develop hypotheses, and design experiments to test them (Johnson and Lawson, 1998; Lawson, 1988). In the best of these laboratories, students answer research questions using online datasets (e.g., genomic sequence data) (Shaffer et al., 2010) or even contribute to such datasets by isolating and characterizing previously undiscovered life forms (e.g., Hanauer et al., 2006). This work can lead to research publications with students as co-authors (e.g., Hatfull et al., 2010).

Although the committee has characterized the strength of the findings as limited, the evidence from biology education research suggests that when compared with traditional laboratory exercises, inquiry-based laboratories can improve students' learning and their short-term retention of biology content (Halme et al., 2006; Lord and Orkwiszewski, 2006; Rissing and Cogan, 2009; Simmons et al., 2008). Inquiry-based laboratories also can improve students' competency with science practices and confidence in their ability to do science (Brickman et al., 2009), and may increase retention of students in the major (Seymour et al., 2004). It is not clear, however, whether inquiry-based laboratories are more effective in dispelling common misconceptions on such topics as the nature of cellular respiration and the origins of plant biomass.

As one example of an inquiry-based laboratory, the Genomics Education Partnership used the Classroom Undergraduate Research Experience and pre- and post-test assessments to evaluate the impact of an authentic *Drosophila* genome annotation project on learning in 472 students at 46 participating institutions (Shaffer et al., 2010). The experimental design allowed for comparisons in knowledge gains between students who identified elements on the genome and engaged in more extensive characterization and students who only identified elements on the genome. For the latter group, pre- and post-test scores were the same. In contrast, the post-test scores of students who engaged in both tasks were nearly twice as high as their pre-test scores. This effort stands out in the biology education research literature because of the scale of the study and the range of institutions involved.

## Engineering

Unique among DBER fields, engineering is an externally accredited practice-based profession. As a result, undergraduate engineering education involves developing technical competencies and preparing graduates for practice (Lynch et al., 2009). Engineering educators are therefore concerned with both affective and cognitive outcomes of laboratory experiences (Feisel and Rosa, 2005). Along these lines, recent efforts to develop inquiry-based engineering laboratories to foster student engagement seem promising (Kanter et al., 2003) although the research is in an early stage of development. However, the committee's review revealed that a limited amount of research exists on how these laboratories affect students' learning. A follow-up paper to a colloquy on the role of laboratory instruction in engineering noted "the lack of coherent learning objectives for laboratories and how this lack has limited the effectiveness of laboratories and hampered meaningful research in this area" (Feisel and Rosa, 2005, p. 121).

## The Geosciences

As with the other fields of DBER, the laboratory is understudied in the geosciences. One study of an introductory geoscience laboratory showed that students who completed the optional laboratory in conjunction with an introductory-level, lecture-based course earned higher final exam scores than students who completed only the lecture course (Nelson et al., 2010). Students over age 25 benefitted much more from the laboratory than students of conventional college age. Older students who took the laboratory option performed 21 percent higher than older students in the lecture-only course, whereas college-age students performed about 3 percent higher than their lecture-only counterparts. Students over age 25 and of conventional college age had similar GPAs and course grades, on average.

## Astronomy

A limited amount of research on the introductory astronomy laboratory suggests that online datasets might have some benefits for undergraduate students. For example, the highly structured task of repeatedly querying large online datasets can enhance students' understanding of the nature of scientific inquiry (Slater, Slater, and Lyons, 2010; Slater, Slater, and Shaner, 2008). In addition, undergraduate students' understanding of the difference between data and evidence can be enhanced when they are explicitly taught to develop their own research questions and conduct investigations over the duration of a course (Lyons, 2011). One study has shown that this approach works equally well for students in face-to-face collaborative groups and individually in the relatively isolated environment of an internet-delivered astronomy course (Sibbernsen, 2010).

## LEARNING IN THE FIELD SETTING

For some disciplines, learning in the field is just as important as learning in the classroom or laboratory. The geoscience curriculum, for example, has had field instruction at its core for more than a century (Mogk and Goodwin, 2012). Field learning in the geosciences encompasses a variety of activities, ranging in scale from a single outdoor class activity (perhaps with a duration of only an hour or two), to sustained individual or group projects, short- or long-term residence programs, capstone field camps at the undergraduate level, and group or individual field projects at the undergraduate or graduate level (Butler, 2008; Mogk and Goodwin, 2012; Whitmeyer, Mogk, and Pyle, 2009).

The geoscience education literature is replete with descriptions of instructional activities in the field. However, reports of the efficacy of these activities are largely observational and anecdotal. We have characterized the strength of this evidence as limited because few studies exist and they have typically been conducted in the context of a single field course. The available research measures a variety of outcomes, and suggests that field courses can positively affect the attitudes, career choices, and lower- and higher-order cognitive skills of student participants as measured by survey instruments designed to assess these outcomes (Huntoon, Bluth, and Kennedy, 2001); improve introductory students' understanding of concepts in the geosciences as measured by the Geoscience Concept Inventory (Elkins and Elkins, 2007); and contribute to the development of teamwork, decision-making, autonomy, and interpersonal skills (Boyle et al., 2004; Stokes and Boyle, 2009). Several scoring rubrics are helping to standardize the assessment of learning outcomes in the field (e.g., Pyle, 2009).

Some studies have used GPS tracking devices to monitor students at work in the field. Building on the cognitive science field of naturalistic decision making (Klein et al., 1993; Lipshitz et al., 2001; Marshall, 1995; Zsambok and Klein, 1997), some geoscience education research has analyzed the navigational choices of students who were engaged in independent field work and correlated those choices with performance (Riggs, Balliet, and Lieder, 2009; Riggs, Lieder, and Balliet, 2009). That research reported an optimum amount of relocation and backtracking in field geology: too much retracing indicates confusion, and too little reoccupation of key areas appears to accompany a failure to recognize important geologic features.

## EFFECTS OF INSTRUCTIONAL STRATEGIES
## ON DIFFERENT STUDENT GROUPS

Most of the studies the committee reviewed were not designed to examine differences in terms of gender, ethnicity, socioeconomic status, or other student characteristics. However, physics education research has explored the impact of instructional innovations on females and minorities. For example, the positive impacts of *SCALE-UP* appear to be even greater for females and minorities (Beichner et al., 2007). In contrast, researchers studying the early implementation of *Workshop Physics* discovered that the attitudes of females about the course were significantly worse than males, and that females' dissatisfaction arose from the alternative format of *Workshop Physics*, difficult laboratory partners, and time demands (Laws, Rosborough, and Poodry, 1999).

Some physics education researchers designed a course called Extended General Physics specifically for students whom they identified as likely to struggle with college physics. Enrollment in the course included nearly 70 percent females, and greater proportions of underrepresented minorities than traditional physics courses. Among other features, the course incorporated several student-centered pedagogies, including collaborative activities. Students in this course had a higher retention rate, higher grades, and better attitudes than their peers in the traditional section, and these differences were particularly pronounced for females and minorities. Moreover, students in Extended General Physics and traditional courses scored similarly on common exam questions, indicating that Extended General Physics was at least as rigorous as the traditional physics course (Etkina et al., 1999).

Along similar lines, a handful of biology education research studies suggest that first-year students from underrepresented groups perform better in biology courses that offer supplemental instruction (Barlow and Villarejo, 2004; Dirks and Cunningham, 2006; Matsui, Lui, and Kane, 2003). This effectiveness might be at least partially attributed to the cooperative learning that is typically included in supplemental instruction (Rath et al., 2007).

A few astronomy education research studies also have examined differences among males and females. One study showed that males outperform females on the *Astronomy Diagnostic Test*, leading the study's authors to conclude that the concept inventories developed for astronomy (see Chapter 4) might have some inherent biases (Brogt et al., 2007; Hufnagel, 2002; Hufnagel et al., 2000). In a separate study, female students in ASTRO 101 started at lower achievement levels than their male counterparts, but the use of curriculum materials designed to improve quantitative reasoning skills closed those initial gaps (Hudgins et al., 2006).

## SUMMARY OF KEY FINDINGS

- *Across the science and engineering disciplines in this study, DBER clearly indicates that student-centered instructional strategies can positively influence students' learning, achievement, and knowledge retention, as compared with traditional instructional methods. DBER does not yet provide evidence on the relative effectiveness of different student-centered strategies, whether different strategies are differentially effective for learning different types of content, or the effectiveness of strategies for subgroups of learners.*
- *Research on the use of various learning technologies suggests that technology can enhance students' learning, retention of knowledge, and attitudes about science learning. However, the presence of learning technologies alone does not improve outcomes. Instead, those outcomes appear to depend on how the technology is used.*
- *Despite the importance of laboratories in undergraduate science and engineering education, their role in student learning has largely gone unexamined. Research on learning in the field setting is similarly sparse.*

## DIRECTIONS FOR FUTURE RESEARCH

Despite the preponderance of DBER on the benefits of student-centered instruction and of instruction that involves the use of technology, important gaps remain. With some exceptions, the studies the committee reviewed measure learning within the context of a single course. Multi-instructor, multi-institutional studies are needed to move beyond the idiosyncrasies of instructional approaches that work well only in the presence of certain instructors or with students who fit a particular profile. More work also is needed on large-scale projects such as POGIL, to better understand the conditions under which its materials are successfully implemented and provide insights into how the effective use of these materials and associated pedagogy can be reliably supported. Additional research examining the influence of student-centered instruction on other types of outcomes, such as declaring a major, retention in the major and pursuing further study also would be helpful. And finally, longitudinal studies are needed to gauge the effects of student-centered instruction on the long-term retention of conceptual knowledge and on the application of foundational skills and knowledge to progressively more challenging tasks.

Most of the research on instructional strategies has been conducted in introductory courses. Less evidence exists regarding the efficacy of different instructional approaches in upper-division courses, although some has been conducted (see, for example, Chasteen and Pollack [2008] and Smith et al. [2011]). Within introductory courses it is unclear whether student-centered learning environments affect different student populations differently, because DBER scholars rarely compare the effects of a given strategy for different student populations. Populations of interest for future study include students who are underrepresented in science, including students for whom English is a second language, females, and ethnic/racial minorities. It also would be useful to explore the dimensions of overall science performance, quantitative skills, and spatial ability. Further study is needed on strategies to accommodate students with disabilities into the full suite of instructional opportunities, especially laboratory and field-based learning.

Across the disciplines in this study, the role of the laboratory class is poorly understood. It would be helpful for scientists, engineers, and DBER scholars to identify the most important outcomes of a well-designed laboratory course, then to design instruction specifically targeted at those outcomes and instruments for routinely assessing those outcomes. Future DBER might compare learning outcomes associated with different types of laboratory instruction (e.g., free-standing versus laboratory activities that are integrated into the main course) and compare outcomes in courses where laboratories are required, optional, or not offered. In addition, laboratory activities in which students conduct inquiry on large, professionally collected data sets (such as genomics data and geoscience datasets served by the U.S. Geological Survey, the National Oceanic and Atmospheric Administration, the National Aeronautics and Space Administration, and various university consortia) have grown in prominence in recent years (Hays et al., 2000), but have been little studied.

Additional research also is needed on field-based learning. Specifically, which types of field activities promote different kinds of learning and which teaching methods are most effective for different audiences, settings, expected learning outcomes, or types of field experiences? The research base is particularly sparse regarding the degree of scaffolding needed for different types of field activities, and which types of field projects are optimal for a given learning goal (Butler, 2008). Given the expense and logistical challenges of field-based instruction, it is important to identify which learning goals (if any) can *only* be achieved through field-based learning, and which (if any) could be achieved through laboratory or computer-based alternatives. These studies also should explore affective dimensions of field learning, including motivations to learn science and cultural and other barriers to learning.

In studying the efficacy of different instructional approaches, DBER scholars must take into account the time constraints of instructors. Future DBER studies might document the time associated with different instructional approaches and explore which approaches are most efficient for supporting students' learning in terms of faculty effort. At the same time, research into enhancing the effectiveness of graduate teaching assistants and paraprofessionals such as full-time laboratory instructors can explore ways to make student-centered instruction an economically viable approach, even at a time of shrinking funding for higher education.

# 7

# Some Emerging Areas of Discipline-Based Education Research

Discipline-based education researchers study several other important aspects of teaching and learning beyond those described in previous chapters. For many of these topics, the research base in discipline-based education research (DBER) is not yet robust. This chapter highlights a few of these topics that are vital to learning science and engineering and warrant further study:

- The role of science and engineering practices in undergraduate education, including in undergraduate research experiences
- Students' ability to apply knowledge in different settings (transfer)
- Students' ability to monitor their own learning processes (metacognition)
- Students' dispositions and motivations to study science and engineering (affective domain)

Some of these topics have been studied by cognitive science researchers or educational psychologists, but they are understudied in DBER for a variety of reasons. They may be addressed implicitly or as a secondary focus in studies on other topics; they may involve basic research (rather than the applied research that dominated the early stages of DBER and remains a strong emphasis today); or they may simply not yet be a priority for DBER scholars. In addition, DBER scholars are just beginning to deploy some of the measurement tools used by scholars in other disciplines that are necessary to research these topics. Despite the relatively sparse DBER literature on these topics thus far, they are of central importance.

This chapter discusses each of the above topics in turn, recognizing the conceptual overlap among some of them and with the topics discussed in Chapters 5 and 6. In contrast to earlier chapters that included a discipline-by-discipline summary of each topic, the evidence base for these four topics does not support such treatment. Instead, we briefly discuss the cross-disciplinary findings—all of which we have characterized as limited because few studies exist, much of the existing research consists of small-scale investigations, and no reviews have been published—and discuss in more detail the findings from the particular DBER field(s) with the most research to date on these emerging topics. Because these topics warrant further study in the context of DBER, each section ends with an identification of directions for future research. However, unlike previous chapters, this chapter does not conclude with a summary of key findings because DBER on these topics is too limited to support conclusions.

## SCIENCE AND ENGINEERING PRACTICES

In part, science may be thought of as a vast and powerful compendium of factual information, concepts, principles, and laws that describe the nature of the universe and its inhabitants. But science also comprises a set of investigative processes, or ways of empirically and systematically studying the natural world, to advance the collective understanding of its order. These investigative processes—which we refer to as practices—and the knowledge gained from their application are critical components of scientific disciplines. Without those investigative practices, there would be no new scientific and engineering knowledge. Thus, an understanding of the attributes of science and engineering practices is vital, as is imparting them to new generations of learners.

In contrast to the clear delineation of content knowledge presented in introductory textbooks, no consensus exists on core disciplinary practices at the undergraduate level. Professional societies emphasize science and engineering practices in different ways. In physics, the American Association of Physics Teachers (1997) provides a set of goals for instructional laboratories that emphasize the central role of practices. In engineering, the ABET accreditation criteria F, G, and H focus on the needed skills of teamwork, communication, and ethics (see Chapter 3). In chemistry, the American Chemical Society Committee on Professional Training revised its guidelines for the training of chemists to include the same skills as engineering.[1]

---

[1]The guidelines are available at http://portal.acs.org/portal/PublicWebSite/about/governance/committees/training/acsapproved/degreeprogram/WPCP_008491 [accessed March 10, 2012].

At the K-12 level, the nature of science has historically received greater attention (Collins and Pinch, 1993; DeBoer, 1991; Petroski, 1996). Indeed, "The idea of science as a set of practices has emerged from the work of historians, philosophers, psychologists and sociologists over the past 60 years" (National Research Council, 2012, p. 43). More recently, *A Framework for K-12 Science Education* identifies core disciplinary ideas, practices, and cross-cutting concepts in the physical, life, and Earth sciences and engineering. That report's conceptualization of practices is useful to consider here (National Research Council, 2012, pp. 44-45):

> One helpful way of understanding the practices of scientists and engineers is to frame them as work that is done in three spheres of activity, as shown in Figure [7-1]. In one sphere, the dominant activity is investigation and empirical inquiry. In the second, the essence of work is the construction of explanations or designs using reasoning, creative thinking, and models. And in the third sphere, the ideas, such as the fit of models and explanations to evidence or the appropriateness of product designs, are analyzed, debated, and evaluated. . . . In all three spheres of activity, scientists and engineers try to use the best available tools to support the task at hand.

The framework goes on to identify eight specific science and engineering practices that advance an understanding of science among students.

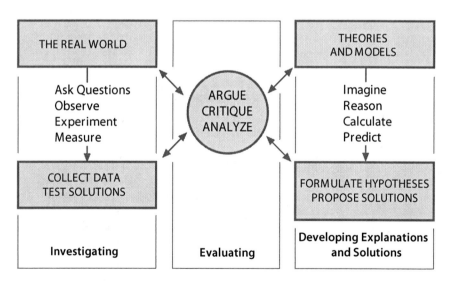

**FIGURE 7-1** The three spheres of activity for scientists and engineers.
SOURCE: National Research Council (2012, p. 45).

These practices also apply to undergraduate education. They include the following (National Research Council, 2012, p. 49):

1. Asking questions and defining problems
2. Developing and using models
3. Planning and carrying out investigations
4. Analyzing and interpreting data
5. Using mathematics and computational thinking
6. Constructing explanations and designing solutions
7. Engaging in argument from evidence
8. Obtaining, evaluating, and communicating information

Learning and becoming adept at science and engineering practices should not be separated from content learning. Rather, research at the K-12 level has shown that well-designed curricula and instructional practices can support deeper learning of content at the same time that students are engaging with these practices (National Research Council, 2007).

## Overview of Discipline-Based Education Research on Science and Engineering Practices

In contrast to K-12 education (see Flick and Lederman, 2004), science and engineering practices at the undergraduate level are largely understudied across the disciplines in this report. Most of the available evidence comes from physics and chemistry, with fewer studies in biology the geosciences, engineering, and astronomy. The studies that are available reflect the considerable range of accepted methods in science, as well as the lack of clear consensus among scientists and engineers about which practices are most important at the undergraduate level.

### Research Focus and Methods

DBER studies of science practices typically adopt one of two perspectives: (1) examining practices as an outcome (i.e., how proficient students are in a specific practice or practices of science and engineering), or (2) engaging in practices as a means of leveraging content learning or other outcomes. Much of the engineering education research on professional skill-related process falls outside of these categories. That research largely describes "how to"—for example, how to build and use teams—rather than studying what works for developing students' professional skills and how those strategies work. However, research on teamwork in particular is shifting toward the latter focus (Svinicki, 2011). Similarly, some notable curriculum development efforts promote practices in biology (e.g., BioQuest, which focuses on problem posing, problem solving, and peer persuasion),[2] although with few efforts to evaluate their efficacy.

Most DBER studies that the committee reviewed on science practices do not explicitly situate themselves in broader theories of learning. An implicit framework for some biology education studies on this topic harkens back to Karplus' (1977) learning cycle as elaborated by Lawson (1988). The learning cycle is fundamentally a constructivist approach, which argues that people generate their own understandings and form meaning as a result of their experiences and ideas (Piaget, 1978; Vygotsky, 1978).

DBER scholars use a range of methods to study science and engineering practices. Evidence is typically derived from self-developed assessments, surveys, interviews, and observations of students in class or laboratory sessions. Five of the 11 studies that the committee reviewed in biology are quasi-experimental studies of students in different courses (Dirks, 2011). The majority of these studies involve only students majoring in the biological sciences, and it is much more common for these studies to take place in lower division courses than upper division courses (Dirks, 2011).

*Students' Proficiency with Practices*

Regarding practices as an outcome, findings from DBER suggest that undergraduate students have little experience or expertise in aspects of designing or conducting scientific investigations that are important to practicing scientists and engineers. Specifically, students struggle to

- distinguish between data and evidence (Lyons, 2011);
- classify matter (Stains and Talanquer, 2008);
- ask scientifically fruitful questions (Karelina and Etkina, 2007; Slater, Slater, and Lyons; 2010; Slater, Slater, and Shaner, 2008); and
- make predictions, observations, and explanations (Kastens, Agrawal, and Liben, 2009; Mattox, Reisner, and Rickey, 2006; Tien, Teichert, and Rickey, 2007).

Some research suggests that students also lack an understanding of experimental uncertainty unless they are explicitly taught about it (see Sere, Journeaux, and Larcher, 1993). These results are consistent with the prevalence of traditional laboratory exercises for introductory students, which are not effective in teaching higher-order skills (Redish, Steinberg, and Saul, 1997).

---

[2]See http://www.bioquest.org for a description of this curriculum [accessed March 31, 2012].

In addition to these studies of general practices, DBER is beginning to generate evidence about specific practices of individual disciplines. In physics, the research on problem solving is extensive, as is research about how students do and do not use discipline-specific models and graphical representations (see Chapter 5 for a discussion of research on problem solving and the use of representations in various disciplines). This research shows that students who use representations outperform those who do not but that students rarely use representations on their own (De Leone and Gire, 2006; Kohl and Finkelstein, 2005; Rosengrant, Etkina, and Van Heuvelen, 2006; Van Heuvelen and Zou, 2001). Research also has shown that some visualization skills that are important to the geosciences can be improved through a targeted set of practice exercises (Titus and Horsman, 2009; see Chapter 5). Also discussed in Chapter 5, engineering education courses that use case analyses, model-eliciting activities, worked out problem examples, and heuristics have been shown to help students develop the practices of problem analysis and design (Svinicki, 2011).

The laboratory is an important setting for students to engage in science and engineering practices, either in the context of regular course work or through research experiences with faculty. In astronomy education research, a limited number of studies address the role of traditional laboratories in improving proficiency with practices, and the results so far are mixed. Although some research suggests that these laboratories do not help students understand that scientists use a wide variety of methods to conduct investigations, they have been shown to help students improve their ability to develop appropriate scientific questions (Slater, Slater, and Lyons, 2010; Slater, Slater, and Shaner, 2008). As discussed in Chapter 6, research on students' experiences in general chemistry (Miller et al., 2004) and analytical chemistry (Malina and Nakhleh, 2003) has found that the goals of the faculty determine the features of the laboratory that students identify as important. Depending on how faculty structure the laboratory experiment and assess student learning, students can perceive of instruments (e.g., spectrophotometers) simply as objects, without any knowledge of their internal workings, or as useful tools for collecting evidence about the behavior of molecules and their properties.

## Using Practices to Enhance Conceptual Understanding

Considerably less research exists on using science and engineering practices to leverage learning. In physics, some studies demonstrate that engaging in scientific practices improves conceptual understanding (Cox and Junkin, 2002; Etkina et al., 2010) and that making predictions can enhance the educational impact of professor-led demonstrations (Crouch et al., 2004; Cummings et al., 1999). In biology, explicit instruction in science practices—either through supplemental instruction or during a regular course—also appears to help students learn content (Coil et al., 2010; Dirks and Cunningham, 2006; Kitchen et al., 2003). For example, at one university, a preparatory short course for first-year students who were not prepared for rigorous coursework in the sciences explicitly taught graphing, experimental design, and science communication. Students who took the short course before taking a biology course earned, on average, higher overall grades in their introductory biology courses than students who did not participate (Coil et al., 2010; Dirks and Cunningham, 2006). Likewise, engineering students identified as having low-spatial ability who were selected to participate in a one-semester preparatory course on spatial visualization skills had better grades in subsequent courses and better retention in the major than similar students who did not participate in the program (Sorby and Baartmans, 2000).

## Research Experiences for Undergraduates

Some colleges and universities use undergraduate research experiences and internships to supplement traditional learning experiences and offer students additional opportunities to engage in the practices of science and engineering outside the course setting. Many undergraduate research experiences are built on the same apprenticeship model as graduate education. These experiences might include opportunities for discovery of new knowledge; rediscovery of knowledge already acquired from mentors or from the larger disciplinary literature, including through replication or simulation of earlier results; and acquiring skills—perhaps doing bench science, learning to use analytical instruments, mastering modeling programs, or doing field work. Research experiences can give the student a sense of whether advanced study and a career in the particular field is a good personal fit (Hunter, Laursen, and Seymour, 2007; Seymour et al., 2004); allow them to experience the social dimensions of the work of science and engineering (Dunbar, 1995); and begin the long process of induction into science and engineering by involving them in a community of practice that shares goals, values, assumptions, and methods (Mogk and Goodwin, 2012; Zuckerman, 1992).

Studies on undergraduate research experiences generally cut across science disciplinary boundaries (e.g., Hunter, Laursen, and Seymour, 2007; Kardash, 2000; Lopatto, 2010; Reuckert, 2002; Sadler et al., 2010). These studies show that students who participate in undergraduate research believe that they have enhanced their research skills and report being more motivated to pursue a career in science after the experience. Other benefits include improved attitudes or dispositions toward science and a better understanding of the nature of science (Jarrett and Burnley, 2010).

A widespread assumption is that extended research experiences will promote more robust knowledge of science content and understandings of scientific ideas and principles, but this assumption has not been adequately tested or borne out (Sadler et al., 2010). The typical methodology is students' or mentors' self-reports via survey or interview; direct assessment of students' pre- and post-apprenticeship knowledge/understanding is rare.

The few studies that have examined group differences show that research experiences enhanced retention in science for students from underrepresented groups (Gregerman, 2008; Locks and Gregerman, 2008; Nagda et al., 1998; Russell, Hancock, and McCullough, 2007), and in some cases, improved academic performance (Gilligan et al., 2007). The most promising findings come from a longitudinal study of the Undergraduate Research Opportunity Program (UROP), which includes many science disciplines and has been in existence at the University of Michigan since 1988. In that research, 75 percent of the African American men who participated in UROP completed their degrees, compared with 56 percent of a control group who applied but were not accepted into the program (Gregerman, 2008; Locks and Gregerman, 2008).

Although research experiences appear to improve student retention, they do not appear to affect students' decisions about their future courses of study. In a multi-institutional survey study of the benefits of research for undergraduates and the progression rate of those students to advanced degrees,[3] 83 percent of the 1,135 respondents earned, or intended to go on to earn, advanced degrees (Hunter, Laursen, and Seymour, 2007; Lopatto, 2007). The percentages of underrepresented minorities and whites who intended to earn advanced degrees were similar. However, the research experiences did not change students' minds about pursuing future study. Along similar lines, a separate study of 51 students found that an undergraduate research program made little difference in the intent of females to pursue a graduate degree in astronomy (Slater, 2010).

It is important to note that most undergraduate research experiences are voluntary. The self-selection of students into these programs potentially confounds the research findings because the students who opt to participate might be more motivated or inclined to pursue further study or a career in science or engineering. Notable efforts to counter self-selection bias in the research include the evaluation of UROP (Gregerman, 2008), which compared students in the program to students who applied but were not

---

[3]Lopatto (2007) validated an instrument to measure how students experience undergraduate research (Survey of the Undergraduate Research Experience, SURE). The quality of the instrument adds weight to the evidence that undergraduate research enhances student perceptions of their science skills and interest in science.

accepted, and a review of four undergraduate programs that compared students and faculty who were involved in the programs and students and faculty who were not involved (Hunter, Laursen, and Seymour, 2007).

### Directions for Future Research on
### Science and Engineering Practices

For most of the disciplines examined in this report, the research on how students learn in laboratories or in the field—where they are likely to engage in science and engineering practices—is scarce. Additional research is needed to better understand how to measure and promote proficiency with these practices, and to explore relationships among practices and other outcomes such as overall understanding of concepts, practices, and ways of thinking of science and engineering.

In addition to the general practices described in this chapter, which span all of the science and engineering disciplines, individual disciplines may emphasize other practices or nuances of the general practices. Future research on practices at the undergraduate level might involve scientists, engineers, and/or scholars of the nature of science in reflecting on such discipline-specific variations and taking them into account when studying student learning.

More specifically, given the increase in undergraduate research programs and the expense of these programs in both time and money, it is important to understand the short- and long-term impacts of undergraduate research experiences and other research apprenticeships. Most reports on research apprenticeships document general trends (e.g., experiences have a positive effect on a specified outcome), but do not investigate the processes or mechanisms by which the outcome is achieved (Sadler et al., 2010). Ideally, future research examining these mechanisms would be conducted in ways that minimize or account for the effects of the self-selection bias of undergraduate research experiences. Additional studies are also needed on research experiences that occur during the regular school year, as most of the published research on the impact of undergraduate research experiences has been conducted on 10-week summer research apprenticeships rather than ongoing, independent, or mentored research in faculty laboratories.

It would be useful to study a wider variety of opportunities that engage students in science and engineering practices (e.g., internships in government or industry settings, service learning experiences, or museum and planetarium programs) and that emphasize different dimensions of practice. As one example of the latter, although knowledge of professional ethics is mandated through the ABET engineering accreditation criteria, engineering education research on enhancing student knowledge of professional ethics is scant, and represents a promising area for future study.

## APPLYING KNOWLEDGE IN
## DIFFERENT SETTINGS (TRANSFER)

Instructors expect that students will be able to apply what they learn in the classroom to new situations encountered inside and outside the classroom. Indeed, a prime goal and fundamental assumption of education is the transfer of knowledge from one context to another. Elementary school students are expected to transfer the subtraction skills learned in the classroom to the problem of making change at a neighborhood lemonade stand. At the college level, if students learn how to apply Newton's second law of motion to a problem involving a block on an inclined plane, they are expected to recognize the applicability of that law in understanding the data collected in a physics laboratory or in designing a device for a mechanical engineering class. Indeed, the idea of knowledge transfer is inherent in the discipline of physics because it is assumed that a very small set of fundamental ideas can be used to explain the diversity of the universe. As a result, physics curricula are driven by the assumption that students will be able to apply what they learn across the physics curriculum and to other science and engineering courses. Geoscience curricula, on the other hand, are rich in opportunities for the transfer and application of concepts and insights *from* physics, chemistry, and biology to the Earth system—such as when concepts from chemistry are applied in mineralogy or concepts from physics are applied in structural geology.

Regardless of the discipline, if students were only able to use what they have learned in exactly the conditions under which the learning occurred, that learning would have little practical value. However, the application of knowledge in different contexts is limited by students' understanding of the conditions under which knowledge applies. That understanding, in turn, typically stays fixed in the domain in which students initially learn the content (Bassok, 2003). For example, a student who learns an equation in physics may have difficulty seeing the applicability of the same equation in algebra. Thus, one of the enduring problems in education—no less an issue at the undergraduate level than in K-12 education—is that transfer of learned material to new situations is much more difficult than educators expect.

### Discipline-Based Education Research on Transfer

*Research Focus and Methods*

Transfer is a two-part process: transfer *during* learning (effect of past learning on new knowledge acquisition) and transfer *of* learning (the degree to which the new learning is applied in future situations) (Sousa, 2011). DBER scholars have partially addressed the first dimension through their extensive exploration of students' conceptual understanding (see Chapter 4). Indeed, much of that research is predicated on the assumption that instructors need to know what their students already know, because prior knowledge can either interfere with new learning (negative transfer) or facilitate it (positive transfer) (Ausubel, 1968; Ausubel, Novak, and Hanesian, 1978; Bretz, 2001; Novak, Gowin, and Kahle, 1984). However, only a small amount of DBER has explicitly focused on either transfer during learning or transfer of learning. In contrast, a considerable number of cognitive science studies on physics problem solving (Bassok, 1990; Bassok and Holyoak, 1989) and mathematical problem solving (Bassok, 2003; Catrambone, 1998; Kaminsky and Sloutsky, 2012; Novick and Holyoak, 1991; Reed, 1987) have investigated transfer.

More DBER studies might be identified post hoc as exploring one or both of these dimensions of transfer. Studies on analogical reasoning (Jee et al., 2010; Sibley, 2009), concept mapping (Clark and James, 2004), the application of skills to novel situations such as the use of the petrographic microscope to identify minerals and then to interpret textures (Gunter, 2004; Milliken et al., 2003), and the relationships between learning in the classroom, laboratory, and field (Mogk and Goodwin, 2012) could provide early insights into transfer.

Much of the physics research on transfer in the context of problem solving is based on the theoretical underpinnings of information processing (Simon, 1978). However, the meaning and even the utility of the idea of knowledge transfer as a theoretical construct is controversial within physics education research (Mestre, 2005). Some scholars attempt to find the basic building blocks of physics knowledge and the mechanism for their interaction. This type of theoretical framework is expressed, for example, in terms of mental resources that are activated by situational perception (Hammer et al., 2005). Scholars working within this approach study building blocks of different scales. Within this theoretical framework, transfer is not a process but the outcome of interactions among building blocks. Knowledge transfer, in this approach, is a derived construct.

Other physics education researchers take a more phenomenological approach, driven by the assumption that learning is so complex that a linear set of mechanisms and interactions, even if they exist, are not adequate to describe knowing. Some of these theoretical frameworks are concerned with specifying meaningful characterizations of transfer (Schwartz, Bransford, and Sears, 2005). Others are concerned with characterizing teaching systems that achieve transfer without ascribing the underlying mental mechanisms (Brown, Collins, and Duguid, 1989).

In chemistry, researchers have analyzed students' transfer of knowledge related to the characteristics and behavior of NaCl and NaCl (aq) (Kelly,

2007; Kelly and Jones, 2008; Teichert et al., 2008), and the transfer of knowledge gained from computer animations designed to help students learn concepts related to the particulate nature of matter (Kelly and Jones, 2008). This research is guided by Ausubel, Novak, and Hanesian's (1978) theory of meaningful learning, which proposes that students store new information according to what they identify as similarities between what they already know and what they need to know. Meaningful learning stands in stark contrast to the strategy of rote memorization, explaining why the latter leaves chemistry students with fragmented pieces of knowledge not useful for building connections to new information (transfer during learning) or in future courses (transfer of learning).

Methods of studying transfer range from individual interviews and experiments in controlled research environments to the analysis of student behavior and written work in classes. Participants typically include students in general introductory courses.

### Students' Difficulties Transferring Knowledge

In DBER, most a priori studies of transfer come from chemistry education research. These studies generally suggest that students have trouble applying knowledge in a new context. A series of seminal studies in chemistry (Nakhleh, 1993; Nakhleh and Mitchell, 1993; Nurrenbern and Pickering, 1987; Pickering, 1990; Sawrey, 1990) have demonstrated that students can memorize how to solve problems that require mathematical manipulations of symbols and chemical formulas, but cannot transfer these skills forward to a similar problem involving particulate images (i.e., drawings of the atoms and molecules). In these studies, students could successfully calculate the mass of products when given information about the reactants and a balanced equation of the reaction. However, when presented with a particulate image of the reactants, they could not correctly identify the particulate image of what remained after the reaction.

A few studies in chemistry have documented students' difficulties identifying the critical attributes of a problem (Kelley and Jones, 2008; Tien, Teichert, and Rickey, 2007), which is an important component of transfer (Sousa, 2011). Those difficulties, in turn, appear to preclude students' abilities to transfer their existing knowledge to a new concept. For example, in one study (Teichert et al., 2008), 19 students were interviewed after completing a laboratory experiment that involved measuring the conductivity of sodium chloride solutions in water and explaining the results in terms of particulate images. They were asked to predict the conductivities of $NaCl(aq)$, $AgNO_3(aq)$, and the resulting solution upon their mixing, and to draw particulate images to support their reasoning. Eighteen of 19 students correctly drew NaCl(aq) as solvated ions, and 15 of 19 did so for AgNO$_3$(aq).[4] Students were then asked to read a paragraph about boiling point elevation—a property they had not yet learned—which explained that this property depended solely upon the number of particles in solution and not their chemical identity. In this context, only 10 of the 19 students drew NaCl(aq) as solvated ions. When the 9 who responded incorrectly were asked to look back at their drawing from the first interview, only 6 of them altered their drawing to reflect solvated ions. Consistent with findings discussed in Chapter 6, these findings suggest that students had difficulty identifying the critical attributes of the problems at hand.

It is not surprising that if novices focus on superficial rather than structural features of problems, their ability to apply learned solution features to new situations will be limited (see Novick, 1988, for relevant experimental evidence from studies of mathematical problem solving). Consistent with research from cognitive science (National Research Council, 1999), the findings of Teichert et al. (2008) suggest that students need help to understand which aspects of problems are critical for determining the appropriate solution method and which are not.

## Directions for Future Research on Transfer

In science and engineering, educators and researchers need a greater understanding of how to widely assess and promote the transfer of knowledge and skills within courses; across courses in the major; and among majors and nonmajors as they take different science courses. This understanding is vital because the same students often take chemistry, physics, and biology courses, and each instructor assumes a certain level of prior knowledge. Longitudinal studies would offer the opportunity to develop measures and use them to carefully document and track instances of negative and positive transfer for multiple cohorts of students. Ideally, these studies would be interdisciplinary to understand how students' knowledge transfers across the suite of science and engineering courses they take in college.

One of the greatest research challenges will be framing transfer in a way that is both measureable and acceptable to instructors and students. The course structure and the usual assessment of students in those courses tend to channel DBER into the narrowest form of transfer, namely assessing the direct application of something learned to a new context as measured by the number of correct answers on a task. Bransford and Schwartz (1999) have developed a broader theory of transfer that emphasizes preparation for future learning, which could be a useful framework for further research

---

[4]When a substance dissolves in a solvent, its ions spread out and become surrounded by the molecules in the solvent. These ions are then called "solvated ions."

efforts in this area. For example, college students were more successful than precollege students when presented with the unfamiliar challenge of designing a recovery program for baby eagles, although neither group was completely successful. Bransford and Schwartz argue that the college students had greater general education experience and that they were able to transfer this experience to the new learning situation—evidence of positive transfer that might be missed using more traditional definitions.

Cognitive science research has illuminated some factors that influence transfer, including the quality and context of original learning; the similarity of problems across settings; and, as discussed, students' recognition of the critical attributes of problems (Bassok, 2003; Sousa, 2011). Although this research has been conducted primarily in the context of investigating analogical problem solving in mathematics and physics, DBER scholars in other disciplines might be able to use this research to design studies that advance the understanding of transfer in different disciplines and learning environments.

Cognitive science research also provides insight into specific conditions under which successful transfer occurs, such as when instructors teach students to monitor their own thinking and learning processes (Bransford and Schwartz, 1999; see discussion of metacognition below) or, in the context of problem solving, when source and target problems are superficially and structurally similar (e.g., Chi and Bassok, 1989; Holyoak and Koh, 1987; Ross, 1984). These insights also might inform future DBER studies.

## METACOGNITION

No one doubts that cognition is important to the work of the scientist and engineer and to learning in those disciplines. Cognition, a synonym for "thinking," is necessary for understanding science concepts, for applying the methods of science to discover new aspects of the natural world, and for using scientific ideas to solve important problems. What may be less obvious is that cognition must be supplemented with *metacognition*. Metacognition is the mind's ability to monitor and control its own activities (Brown, 1978; Flavell, 1979). Metacognition is "thinking about thinking." It is a form of higher-order cognition that allows individuals to monitor the quality of their thoughts and to redirect their thinking processes as needed. Metacognition also comes into play during problem solving when the individual judges whether the chosen strategy is effective.

Metacognition is a potent intellectual competency. Students need to be metacognitive so that in every learning context—during lecture, independent reading, laboratory work, research, or discussions—they monitor their comprehension and, if comprehension fails, take corrective steps. A metacognitive learner asks: Do I truly understand? Is my strategy working?

When a student is preparing for a midterm or final exam, a crucial question must be posed: Am I ready to take the test?

An extensive research base in psychology indicates that the ability to make an honest and accurate appraisal of one's own knowledge state is crucial to academic success. Differences in metacognitive ability translate to differences in students' learning outcomes (Tobias and Everson, 1996). In general, students who are more metacognitive are better students overall, which suggests that one goal of education should be to help students become more metacognitive (Lin, Schwartz, and Hatano, 1995).

Metacognition is more than the binary distinction of to know or not. Understanding is always incremental. Indeed, the entire scientific enterprise is predicated on a sense of which aspects of the natural world are well understood, which are partially understood, and which are unknown. These metacognitive sensibilities steer the research agenda, alert investigators to promising questions, and give insights into whether investigative probes are answering the driving questions. Thus, metacognition is an essential competency for both learning and knowledge creation.

Metacognitive approaches are embedded in instructional practices such as problem-based learning, knowledge surveys, and reflective exercises during classes, and in activities designed to support critical thinking. Unfortunately, many instructors assume either that undergraduate students already have the requisite metacognitive skills, or that these skills are too advanced to teach in introductory courses (Trigwell et al., 2001; Yerushalmi et al., 2007). However, research has shown the benefits of teaching metacognitive skills as part of learning content (Collins, Brown, and Newman, 1989). Such practices include being explicit with students about the rationale for learner-centered pedagogy, including defining learning objectives, demanding more student responsibility in mastering content, and using class time for problem solving (Heller et al., 1992). The concomitant need is for students to become more aware of their own study habits and how to improve them (e.g., Silverthorn, 2006). (See also the discussion of cognitive apprenticeship in Chapter 5.)

## Discipline-Based Education Research on Metacognition

### Research Focus and Methods

Students' metacognition is an implicit focus of some research on problem solving and other kinds of decision making, and is increasingly an explicit focus of some DBER. Most of the research that the committee reviewed investigates or assesses the role of metacognition in specific learning environments, typically in the context of problem solving (see Chapter 5), and some research focuses on the development of tools to measure metacognition (e.g., Cooper and Sandi-Urena, 2009). A few studies document efforts to develop students' metacognitive skills (McCrindle and Christensen, 1995). DBER scholars study metacognition through interviews, case studies, quantitative studies, mixed methods, and phenomenology.

Physics studies on metacognition typically take place in a controlled environment outside of the classroom with a small number of students. Participants in these studies are typically students who know some physics, often paid volunteers who have taken one or two introductory courses in physics. This sort of study often uses interviews or students' self-explanations to analyze students' reasoning processes as they engage in a task. In other disciplines, researchers have assessed metacognitive activity in the context of a learning environment such as an inquiry laboratory, or a specific problem solving activity. In biology, for example, McCrindle and Christensen (1995) conducted an experiment in which they randomly assigned students in their introductory class to either a treatment group or a control group to test a strategy for developing metacognitive skills.

Assessing metacognition during learning can be challenging because it is largely a hidden skill, although there are techniques to infer its existence. Moreover, asking students to document metacognitive activities might artificially prompt metacognitive behavior. Such behavior is desirable, but the false positive results could confound the research findings. In chemistry, the Metacognitive Activities Instrument (MCAI), a validated self-report instrument, probes students' thinking about problem solving (Cooper and Sandi-Urena, 2009). As described in Box 7-1, students' MCAI scores have been shown to correlate with their problem-solving strategies and abilities as measured by the online Interactive Multi-Media Exercises system (Cooper, Sandi-Urena, and Stevens, 2008). Besides the MCAI, evidence of metacognition in chemistry has been gathered from self-reports and observations of student work samples and behavior.

*Promoting Metacognition to Enhance Learning*

The effectiveness of deeper, more meaningful processing for retention of information was first documented in cognitive psychology by Craik and Lockhart (1972). Consistent with findings from cognitive science research (National Research Council, 1999), DBER suggests that students can develop metacognitive skills over time when metacognitive strategies are built into instruction (McCrindle and Christensen, 1995; Weinstein, Husman, and Dierking, 2000), but that relatively few students report using metacognitive strategies such as self-testing when studying on their own (Karpicke, Butler, and Roediger, 2009).

One focus of DBER on metacognition concerns the self-explanation effect, which is the benefit to learning and problem solving that accrues

---

**BOX 7-1**
**Chemistry Education Research**
**Metacognition and Problem Solving**

The Metacognitive Activities Inventory (MCAI) was developed by Cooper and Sandi-Urena (2009) to measure students' monitoring of their own thinking during problem solving. In prior work, the researchers investigated student learning using the Interactive Multi-Media Exercises system (Cooper et al., 2008; see Chapter 5). Through the use of hidden Markov modeling and artificial neural networks, Cooper and colleagues identified two dimensions of metacognition: the strategy state (the metacognition involved for a particular solution) and the ability state (a measure of problem difficulty). Students' problem-solving strategies and abilities were significantly correlated with MCAI scores (Cooper, Sandi-Urena, and Stevens, 2008; Sandi-Urena, Cooper, and Stevens, 2011).

Because the MCAI is a self-report and the IMMEX scores are derived from data mining of student performance, the combination of methods allows for triangulation to assess different interventions. For example, students who participated in a workshop designed to promote meta-cognition showed consistent changes in their MCAI scores (Sandi-Urena, Cooper, and Stevens, 2011), and students who take laboratories designed to prompt metacognitive activity also show significant gains (Sandi-Urena, Cooper, and Stevens, 2012).

---

from attempts by learners to explain to themselves or another person concepts that are unclear. For example, studies of how college students use example solutions provided in physics textbooks to learn topics in mechanics found that successful students explained and justified solution steps to a greater extent than did unsuccessful students (Chi et al., 1989). The quality of the explanations also differed; good students referred to general principles, concepts, or procedures they had read about in an earlier part of the text and examined how they were instantiated in the current example (Chi and VanLehn, 1991).

Replication of these studies in biology (Ainsworth and Loizou, 2003) has yielded similar results. This research examined whether the way information about the human circulatory system is presented affects learners' self-explanations and subsequent learning. The learners were college students who had not taken a biology class beyond high school. They took a pre-test, learned about the circulatory system via either text or diagrams, and then took a post-test. Students were directed to generate explanations to themselves during the study phase, and the students who generated more self-explanations did better on the post-test. In addition, students who received diagrams provided more self-explanations, performed better on the post-test, and spent less time studying the material than students who received text. Similarly, research in engineering education has found that incorporating metacognitive reflection steps and self-explanation prompts into instruction can improve students' problem solving (Svinicki, 2011).

Together, this research suggests that generating and articulating explanations can be an effective pedagogical tool to help students process more deeply the underlying structure of unfamiliar concepts and problems in all disciplines. This deeper processing can lead to enhanced learning compared with what is achieved by simply reading textbook paragraphs and examples or examining textbook diagrams.

### Directions for Future Research on Metacognition

Metacognition is a necessary skill for meaningful learning and thus merits continued study in the context of DBER. Further research could clarify which metacognitive skills are useful to science and engineering. Because the skills may not be the same for each discipline, additional DBER could examine these similarities and differences. Other efforts might be directed at developing easy-to-use assessments that align to the appropriate metacognitive skills and that measure the effects on metacognition of instructional interventions and learning environments. Finally, the kinds of conditions and strategies that limit or promote metacognition (e.g., writing across the curriculum, structured problem solving, and assessments that promote metacognition) also merit examination. In this regard, the MORE model (model, observe, reflect, explain) (Tien, Rickey, and Stacy, 1999), which promotes metacognition in the chemistry laboratory, might be adapted to other settings and disciplines. The spread of cooperative group problem solving from physics to other disciplines such as chemistry, engineering, and health sciences also might be instructive (Heller, Keith, and Anderson, 1992). This approach emphasizes the practice of metacognitive skills by making them evident through the social interaction of co-constructing a problem solution with the use of specific scaffolding.

## DISPOSITIONS AND MOTIVATION TO STUDY SCIENCE AND ENGINEERING (THE AFFECTIVE DOMAIN)

Successful science and engineering education cannot be defined solely in terms of how many concepts and practices students learn. Students have attitudes, beliefs, and expectations about learning that can influence their behavior and performance in courses (Halloun, 1997; Hammer, 1994, 1995; May and Etkina, 2002; Perkins et al., 2005). As an example, a common student belief is that physics consists of many unrelated pieces of information. This belief often leads students to approach physics by memorizing formulas without connecting them to broader, underlying concepts and principles (Hammer, 1994, 1995). More generally, helping students become members of a scientific or engineering community requires attention to a wider range of outcomes and how to achieve them.

Educational outcomes beyond the mastery of discipline-specific content are considered part of the affective realm (Snow, Corno, and Jackson, 1996). The affective domain is very broad. Psychologists use the term "affect" to refer to an observable expression of emotion, but the term can extend to motivation, attribution, and willful action (i.e., volition). Affect can also refer broadly to belief systems, including the important concept of self-efficacy (Bandura, 1986), or the belief that one is capable of accomplishing a goal such as learning within a particular discipline. The affective domain also can be defined to include a range of external and internal factors that influence a student's ability to learn, including values, social pressures, stereotypes, perceptions, feelings, anxiety, or fear (Krathwohl, Bloom, and Masia, 1964). Another important concept is conation, which refers to volition, or more generally, to the connection between knowledge and affect (personal will, motivation) and intentional, goal-oriented personal actions or behaviors (the desire to learn more, or to engage in proactive activities). Conation provides critical evidence of self-direction and regulation (Hilgard, 2006; Snow, 1989). All such feelings, motives, and beliefs fall under the banner of affect or the affective domain as we use those terms here.

Scholars are increasingly aware that affective aspects of learning are directly linked to cognitive and memory functions (Gray, 2004; Pessoa, 2008; Storbeck and Clore, 2007). Indeed, sociocultural perspectives on learning recognize that a changing sense of one's own identity and competence in a domain occurs at the same time that one is becoming increasingly adept at disciplinary practices and knowledge application and that these concurrent processes are mutually supportive (Lave and Wenger, 1991; Resnick and Klopfer, 1989; Rogoff and Wertsch, 1984). These findings suggest that researchers and instructors should not consider cognitive and affective development apart from each other.

### Discipline-Based Education Research on the Affective Domain

The Carnegie Preparation for the Professions Program (Sullivan, 2005) describes three apprenticeships: the apprenticeship of the head (intellectual development), the hand (skill development), and the heart (development of habits of mind, values and attitudes). Many science disciplines and engineering stress the first apprenticeship (the head), place less emphasis on the second (the hand), and are silent or implicit about the third apprenticeship (the heart) (Sullivan, 2005). Similarly, in DBER, the affective domain has received less attention than the cognitive domain.

## Research Focus and Methods

The research that has been done in DBER is as broad as the affective domain itself, and ranges from students' views about the discipline, to their motivations for pursuing science and engineering, to the social dimensions of fieldwork, to the role of student beliefs in conceptual change.

Scholarly research on the affective domain rigorously probes students' attitudes and beliefs about content, pedagogy, the discipline as a whole, and/or learning in general. DBER scholars use a range of instruments to measure aspects of the affective domain. Some widely used, validated instruments include the following:

- Epistemological Beliefs Assessment for Physical Sciences (Elby, 2001), which assesses students' views about the nature of knowledge and learning in physics.
- Science Motivation Questionnaire, which assesses six components of motivations: (1) intrinsically motivated science learning, (2) extrinsically motivated science learning, (3) relevance of learning science to personal goals, (4) responsibility (self-determination) for learning science, (5) confidence (self-efficacy) in learning science, and (6) anxiety about science assessment (Glynn and Koballa, 2006).
- Colorado Learning Attitudes about Science Survey (CLASS), which is designed to compare novice and expert perceptions about the content and structure of a specific discipline, the source of knowledge about that discipline, and the connection of a discipline to the real world (Adams et al., 2006). Discipline-specific versions of the CLASS exist for physics, (Adams et al., 2006), biology (Semsar et al., 2011), and chemistry (Barbera et al., 2008).

Insight into student and faculty attitudes and beliefs has also come from in-depth interviews and case studies of a few individuals.

## Students' Attitudes, Beliefs, and Motivations

Although the evidence is limited, DBER on the affective domain suggests that student attitudes about physics, chemistry, biology—and, indeed, science in general—differ markedly from the views of practicing scientists. Student attitudes in physics after instruction diverge further from "expert-like" norms than before instruction, even when instruction is more student-centered (Adams et al., 2006; Kost-Smith, Pollock, and Finkelstein, 2010; Redish, Steinberg, and Saul, 1998). In biology as well, students in one study became less expert-like following the introductory course, but subsequently became more expert-like in upper division courses (Semsar et al., 2011). In chemistry, nonmajors' attitudes, as measured by an instrument known as CHEMX, move toward faculty norms during the first year of chemistry instruction. Chemistry majors, on the other hand, move away from the faculty's norms during the first year, and then begin moving closer toward faculty views when they take organic chemistry (Grove and Bretz, 2007).

Although motivation is largely understudied in DBER, a longitudinal (7-year) study in engineering found that several factors influence students' motivation to study engineering, including "psychological/personal reasons, a desire to contribute to the social good, financial security, or, in some cases, seeing engineering as a stepping stone to another profession" (Atman et al., 2010, pp. 3-4). That study also demonstrates that motivation is related to important outcomes such as the intention to pursue engineering as a major and a career.

**The Geosciences.** The nature of some subject matter covered in the geosciences makes consideration of the affective domain (e.g., sensory input from a natural setting) particularly important to scholars in that discipline. In addition, many applications of the geosciences also are directed toward controversial issues such as evolution, the age of the Earth, or climate change. Studying these issues may require students to confront prior beliefs and values. (Student beliefs are also of interest to biologists because of the debates surrounding evolution among some nonscientists—see Chapter 4 for a discussion.)

The Geoscience Affective Research Network has conducted research on the affective domain, with an emphasis on the attitudes and motivations of introductory students (McConnell and Kraft, 2011; van der Hoeven Kraft et al., 2011). In a composite study of introductory classes (7 colleges, 14 instructors, 800 students), student performance was most strongly correlated with scores on the self-efficacy section of the Motivated Strategies for Learning Questionnaire (Pintrich and DeGroot, 1990). In addition, students with low Geoscience Concept Inventory (GCI) scores or low incoming grade point average (GPA) but high self-efficacy earned the same grade as students with high GCI scores or high GPA and low self-efficacy (McConnell et al., 2009, 2010).

Geoscience education research also has documented differences in attitudes and self-efficacy among males and females. One study of 539 males and 607 females from 14 introductory classes at 7 institutions used the Motivated Strategies for Learning Questionnaire to measure pre- and post-course attitudes (Vislova et al., 2010). On the pre-test, females reported lower self-efficacy and higher test anxiety than males. On the post-test, females reported lower likelihood of engaging in future geoscience courses, despite earning similar course grades as their male peers. In a different study, Liben, Kastens, and Christensen (2011) found that female undergraduates' self-reported confidence in the quality of their performance on strike-and-dip and direction tasks was lower than their actual performance or their general spatial ability.

Learning in the field setting, which is an integral part of geoscience education, also has a strong affective component. Fieldwork can engage a wide spectrum of students in learning, in part because of the social interaction it entails (Boyle et al., 2007; Fuller et al., 2006; Maguire, 1998; Marques, Praia, and Kempa, 2003; Stokes and Boyle, 2009). Social aspects of learning in the field include heightened interpersonal interactions, building friendships, and reducing social barriers (Crompton and Sellar, 1981; Fuller et al., 2006; Fuller, Gaskin, and Scott, 2003; Kempa and Orion, 1996; Kern and Carpenter, 1984; Tal, 2001). Well-designed field experiences are seen as an effective means to recruit students to Earth science majors (Karabinos, Stoll, and Fox, 1992; Kern and Carpenter, 1984, 1986; Manner, 1995; McKenzie, Utgard, and Lisowski, 1986; Salter, 2001) and to introduce nontraditional students to the geosciences (Elkins, Elkins, and Hemmings, 2008; Gawel and Greengrove, 2005; Semken, 2005). In addition, some studies have shown that student attitudes toward the geosciences—and indeed, science in general—become increasingly positive as a result of fieldwork (Huntoon, Bluth, and Kennedy, 2001; Stokes and Boyle, 2009), perhaps because students view learning in the field as more interesting than learning in other contexts (Maguire, 1998; Stokes and Boyle, 2009).

### Directions for Future Research on the Affective Domain

To date, much DBER has treated cognitive and affective outcomes as distinct "variables." Future DBER on the affective domain should avoid this dichotomy and recognize the interdependence of affect and cognitive outcomes.

Instructors and researchers would benefit from a greater understanding of the attitudes and beliefs that are the most salient to learning science and engineering, including the role of cultural and social factors and potential differences among different groups of students (e.g., Brandriet et al., 2011). Cognitive science can help DBER scholars to clarify distinctions in theories of affect as they apply to student learning. Such distinctions are useful because they offer new ways to think about undergraduate science education. To that end, research on the affective dimensions of K-12 science learning (e.g., Simpson et al., 1994) also might be applied to DBER.

Systematic research on student motivation in the sciences and engineering is lacking. Future DBER studies might build on the extensive literature on motivation, especially expectancy and value orientation in cognitive science and in the broader higher education literature (e.g., Ambrose et al., 2010; Svinicki, 2004).

Research on a range of teaching strategies that engage the affective domain (e.g., collaborative study; teaching controversial issues; human impacts of course content) and that have the potential to change student attitudes and beliefs also would be useful (e.g., Middlecamp, 2008). In this realm, the interplay between faculty behavior and students' affect merits further exploration: Do faculty responses to student reactions influence teaching strategies and, as a result, student learning? Instructors also have attitudes, beliefs, and values about students and how they learn. The complex interaction of these elements influences how and what instructors teach. Thus, the attitudes and beliefs of instructors themselves should be studied to understand their expectations for student learning in science and engineering—perhaps building on work that has already been conducted in physics (Geortzen, Sherr, and Elby, 2009, 2010; Henderson and Dancy, 2007; Henderson et al., 2004, 2007; Yerushalmi et al., 2007).

On a broader level, research on multiple dimensions of the affective domain would enhance the understanding of "what works" in the recruitment and retention of students into science and engineering majors, with longitudinal studies to determine which career paths students ultimately choose (e.g., Connor, 2009). As one example, in light of the larger percentage of undergraduate females majoring in biology compared to the physical sciences, studies that focus on the persistence of females in undergraduate majors and careers in the life sciences would be illuminating.

# Section III

# *Future Directions for Discipline-Based Education Research*

# 8

# Translating Research into Teaching Practice: The Influence of Discipline-Based Education Research on Undergraduate Science and Engineering Instruction

One of the long-term goals of discipline-based education research (DBER) is to contribute to the knowledge base in a way that can guide the translation of DBER findings to classroom practice (see Chapter 1). To examine the translation of DBER into instructional practice, the committee was charged with two related questions:

1. To what extent and how has DBER informed teaching and learning in the various disciplines?
2. What factors are influencing differences in the state of research and its impact in the various disciplines?

As discussed in Chapters 4 through 6, DBER scholars often develop research-based teaching strategies and test those strategies in their own classes. Clearly, DBER has informed teaching practice in these classes, with demonstrated gains in student learning in many cases. However, within each science or engineering discipline, DBER scholars comprise only a small fraction of all faculty members. We therefore address the two questions above by examining the extent to which DBER has informed the teaching of science and engineering faculty members who are *not* DBER scholars.

In education as in other fields, translating research into practice has posed a challenge for decades, and many have argued that efforts to integrate the two have met with limited success (Feuer, Towne, and Shavelson, 2002). Two types of translation are (1) translating basic research into interventions or programs, and (2) translating interventions that have had localized success into larger-scale interventions. Both are relevant to DBER.

Regardless of which dimension of translation is the goal, determining the extent to which DBER has informed teaching practice is difficult for many reasons. First, there is a limited empirical baseline of faculty members' instructional practices in science and engineering—few studies have rigorously examined instructional practices within disciplines, and even fewer have studied practices across disciplines at the undergraduate level. Second, because faculty members may draw on similar findings from DBER, cognitive science, educational psychology, science education, education, and/or the scholarship of teaching and learning to inform their practice, it is difficult to disentangle the effects of DBER from those of related research. Third, DBER and related research can influence teaching practices to varying degrees, from increased awareness of student learning challenges to complete transformation of instructional approaches. It is difficult to measure some of the more indirect effects, such as increased awareness, on instruction. And finally, as research on higher education policy and organization has shown, instructional decisions—including the decision to incorporate DBER and other research—are influenced by many more factors than the mere availability of research (Fairweather, 2008). For example, science and engineering faculty are likely to be concerned with fitting new techniques into their overall teaching, research, and service responsibilities. Factors including rewards, the relative importance of teaching and research, and an institutional emphasis on bringing in research money are major influences on these decisions (Austin, 2011). However, research on the importance of these factors relative to each other and the ways in which they interact to influence instruction is relatively scarce (see Quinn-Patton, 2010).

This chapter discusses the available national research on current teaching practices; describes research on efforts within the sciences and engineering to increase faculty members' adoption of research-based practices, including findings from DBER; and situates those efforts in the broader context of research on the factors that influence faculty members' instructional decisions and change in higher education institutions. The committee recognizes that a wide variety of practices affect student learning, such as advising, co-curricular learning activities, and learning communities comprised of faculty members or students. However, the focus of this chapter is on classroom practices.

## THE CURRENT STATE OF TEACHING IN UNDERGRADUATE SCIENCE AND ENGINEERING

The documentation of instructional practices is based on two types of sources. First are national surveys of faculty work that have a representative sample of disciplines, including science and engineering. These surveys include the National Surveys of Postsecondary Faculty sponsored by the National Center for Education Statistics (e.g., Schuster and Finkelstein, 2006; U.S. Department of Education, 2005) and the Higher Education Research Institute surveys sponsored by the University of California, Los Angeles (e.g., DeAngelo et al., 2009). Second are a limited number of national surveys of faculty members in engineering (Borrego, Froyd, and Hall, 2010), the geosciences (Macdonald et al., 2005), and physics (Henderson and Dancy, 2009).

The sample sizes of national surveys of faculty permit cross-disciplinary comparisons only at a gross level such as natural sciences versus social sciences. The discipline-specific surveys, in contrast, contain larger samples in the disciplines studied. However, those results are not readily comparable across the disciplines because researchers asked different types of respondents about different types of research-based practices. In addition, findings from the individual disciplinary surveys must be interpreted with caution. Response rates vary from 12 percent in engineering to 50 percent in physics (compared with almost 90 percent across all disciplines in the National Surveys of Postsecondary Faculty 1994). The results may reflect selection bias, if faculty members more engaged in research-based teaching responded more frequently than others who were less engaged. In addition, the findings in the geosciences and physics are based on faculty self-reports, which may overestimate the extent of change in teaching practice. The engineering surveys report on department chairs' perceptions of faculty members' teaching practices, which also might not be accurate.

National survey results of faculty instructional approaches show that faculty members in science and engineering fields are, on average, the least likely to use any form of student-centered or collaborative instruction. They are the most likely to rely primarily on lectures in their classrooms (Fairweather, 2005; Fairweather and Paulson, 1996, 2005; Schuster and Finkelstein, 2006). These results are consistent with the more detailed studies of individual science and engineering disciplines described next.

Researchers in the geosciences conducted a web-based, national survey of 2,207 faculty members in 2004, with a 39 percent response rate (Macdonald et al., 2005). Survey responses revealed that traditional lecture was the most commonly used classroom teaching method. Sixty-six percent of those teaching introductory classes and 56 percent of those teaching courses for majors reported lecturing in nearly every class. More than half of respondents said they incorporated some interactive activities at least once a week, usually lecture with questions or lecture with demonstrations. Faculty reported using interactive techniques more frequently in courses with fewer than 31 students—including small introductory courses and courses for majors—than in medium-sized (31-80 students) or large classes (more than 80 students).

The extent to which instructors in the geosciences reported using interactive activities varied. Most respondents reported asking students to solve problems and analyze data, although they rarely asked students to pose and solve their own problems. In addition, a majority of respondents teaching introductory courses and 80 percent of those teaching courses for majors, asked students to read journal articles at least once per semester. In 2009, these data were used as the baseline in an evaluation of On the Cutting Edge (see "Efforts to Promote Research-Based Practices in Science and Engineering" in this chapter for a discussion of the evaluation).

Taking a slightly different approach, surveys of teaching practice in physics and engineering[1] focused on the diffusion of innovations, using Rogers' (2003) theory of the process involved in adopting an innovation (see Box 8-1) as a frame for data collection and analysis. In a nationally representative sample of 722 physics faculty across the United States (with a 50 percent response rate), nearly all respondents (87 percent) indicated familiarity with one or more research-based instructional practice(s) and approximately half (48 percent) reported currently using at least one such practice (Henderson and Dancy, 2009). At the same time, faculty reported they frequently modified the research-based practices (40 percent reported minor modifications, 41 percent reported major modifications, and only 17 percent reported implementing the research-based practice with fidelity to the developer's design). In addition, faculty frequently discontinued a practice, typically after trying it for at least one semester. The reported rate of dropping a practice ranged from 30 to 80 percent, depending on the practice, with an overall average of 40 to 50 percent.

Based on these results, Henderson and Dancy (2009) argue that current physics education research dissemination approaches (such as journal articles, conferences, and workshops) have been more successful in raising widespread awareness of new instructional practices than in helping faculty understand the underlying principles of these practices, or how to deploy them effectively. Moreover, Henderson and Dancy suggest that the high level of discontinuance (even after modification) indicates that faculty either lacked the knowledge needed to customize a research-based practice to their local situation or underestimated the factors that tend to work against the use of innovative instructional practices. Indeed, Rogers (2003) warns that a lack of knowledge of how to implement the innovation correctly or of its underlying principles can lead to discontinuance of an innovation. These findings argue for shifting the conversation from "what works" and the

---

[1]Because of the low response rate (12 percent) in the Borrego, Froyd, and Hall (2010) survey of engineering department chairs, and the fact that research universities were overrepresented in the sample compared with the universe of engineering departments, the committee chose not to report results from that study.

---

**BOX 8-1**
**Rogers' (2003) Theory of the Innovation-Decision Process**

As defined by Rogers (2003), the process of deciding to adopt an innovation includes five stages:

Stage 1: Knowledge. The individual learns about the innovation and seeks information about it.

Stage 2: Persuasion. The individual evaluates the innovation and begins to develop a positive or negative attitude. Close peers' evaluation of the innovation have the most credibility.

Stage 3: Decision. The individual decides to adopt or reject the innovation.

Stage 4: Implementation. The individual puts the innovation into practice, possibly with some modifications, yet some uncertainty remains.

Stage 5: Confirmation. The individual looks for support for his or her decision. At this stage, the individual may decide to discontinue the innovation, either by replacement (adopting a better innovation) or by "disenchantment," because the innovation does not meet the individual's needs.

---

concomitant evidence for those practices to putting proven practices into place efficiently.

## EFFORTS TO PROMOTE RESEARCH-BASED PRACTICE IN THE SCIENCES AND ENGINEERING

Efforts to increase the impact of DBER on instruction must be viewed in the broader context of currently proliferating efforts to promote research-based undergraduate instruction in science and engineering. Professional societies, federal funding agencies, and accreditation organizations, all located outside academic institutions, have worked to inform faculty members about research that can inform their teaching and encourage them to change their teaching practices. Not all of these efforts focus solely on DBER, and the extent to which they emphasize DBER is unknown.

The National Science Foundation (NSF) has been an important external force in undergraduate science and engineering education, encouraging faculty to use research—including DBER—to improve their teaching practices.

Financial support from NSF has sponsored conferences and professional development opportunities across the science and engineering disciplines. NSF also funds research into the process of change in science and engineering faculty teaching practices, including individual studies (e.g., Henderson and Dancy, 2009), and national conferences ("Facilitating Change in Science and Engineering Undergraduate Education"[2] in 2008 and "Vision and Change in Undergraduate Biology Education"[3] in 2009 to help improve instructional quality in biology).

Current NSF efforts build on three decades of national discussion and debate about the need to improve science teaching and learning (National Science Board, 1986), including National Research Council reports (2007, 2009) calling for rapid increases in the number and quality of science and engineering graduates if the United States is to remain competitive. In response, NSF now has a clear mission "to support science and engineering education programs at all levels and in all fields of science and engineering."[4]

Professional societies also have played important roles in undergraduate education. In engineering, the accrediting agency ABET is a particularly important external force (see Chapter 2 for a more detailed discussion). Accreditation standards introduced by ABET in the 1990s were aimed at improving the quality of engineering teaching and learning (ABET, 1995). The standards call for an outcomes-focused, evidenced-based cycle of observation, evaluation, and improvement of instruction. ABET reinforces the importance of teaching and learning in engineering by requiring programs seeking accreditation to demonstrate effective instructional practices and learning outcomes. This external organization has increased the commitment of engineering programs to student learning (Lohmann and Froyd, 2010) and led to documented improvements in student learning (Lattuca, Terenzini, and Volkwein, 2006). In a related vein, the American Chemical Society approves programs but does not accredit them. Participation in the approval process is voluntary, and the committee did not find evidence demonstrating the impact of the approval process on chemistry programs or student learning.

Through these and other efforts, science and engineering faculty and future faculty have many options for professional development that is focused on integrating research into practice, ranging from campus-based initiatives to national programs. The following discussion concerns national-level

---

[2]For more information, see http://www.wmich.edu/science/facilitating-change/ [accessed February 17, 2012].

[3]For more information, see http://visionandchange.org/ [accessed March 13, 2012].

[4]For NSF's authorizing language and rules, see http://www.nsf.gov/od/ogc/leg.jsp [accessed March 31, 2012].

initiatives for which research or evaluation data exist. Given the committee's charge to examine the extent to which DBER has informed instruction, the bulk of the discussion addresses discipline-specific professional development opportunities for faculty and future faculty because DBER is most likely to be incorporated into discipline-specific initiatives. DBER also might be used in professional development efforts that span multiple science and engineering disciplines, such as Project Kaleidoscope,[5] but research and evaluations on such programs are not currently available.

The committee recognizes that non discipline-specific professional development that addresses research-based principles of teaching and learning also has the potential to increase the use of discipline-specific, research-based practices such as those identified by DBER. However, the committee found no research establishing linkages between these broader programs and the influence of DBER on instruction. The committee also recognizes that multiple professional development opportunities (as well as other factors) interact to influence faculty members' practices—ranging from increased awareness to actual changes in practice. However, it was beyond the scope of this study to examine the extent to which professional development writ large affects teaching practice or to describe the landscape of professional development activities available to science and engineering faculty. For these reasons, the committee's analysis of the extent to which DBER has informed instruction excluded more general professional development programs.

### Large-Scale, Discipline-Specific Professional Development

Disciplinary and cross-disciplinary societies have implemented national professional development programs and workshops designed to encourage the use of research to change teaching practices. Some notable examples that have been evaluated include the following:

- **The New Faculty Workshop in Physics and Astronomy,** established in 1996 and sponsored by the American Association of Physics Teachers with financial support from NSF and in partnership with the American Physical Society and the American Astronomical Society. These workshops promote research-based reforms that new faculty can adopt with minimal time commitment and minimal risk to their chances of winning tenure (Krane, 2008).
- **The National Academies Summer Institute for Undergraduate Education in Biology,** established in 2003, these week-long intensive

---

[5]For more information on Project Kaleidoscope, see http://www.pkal.org/ [accessed March 31, 2012].

professional development workshops for university faculty emphasize the application of teaching approaches based on education
research, or "scientific teaching" (Handelsman et al., 2004; Wood
and Gentile, 2003a, 2003b).

- Faculty networks and professional development opportunities associated with **POGIL (Process-Oriented Guided Inquiry Learning)**
and PLTL (Peer Led Team Leading), two major initiatives that
are sponsored by NSF and originated from the systemic change in
chemistry initiatives in the 1990s. Specifically, the POGIL Project[6]
offers workshops, classroom and laboratory materials, consulta-
tions with POGIL experts, funding for visits to locations where
POGIL is used, and regional POGIL networks to help teachers
integrate guided inquiry and exploration into the classroom.

- **The National Effective Teaching Institute in Engineering Education,** a multiday workshop established in 1991 to familiarize engineering faculty members with proven, student-centered strategies
(Felder and Brent, 2010).

- **On the Cutting Edge,** a professional development series that
includes workshops and a related website dedicated to content
knowledge and teaching strategies in the geosciences (Macdonald
et al., 2004; Manduca, Mogk, and Stillings, 2004).

The extent to which these programs are based on DBER relative to
other, related research has not been systematically documented. As a result,
no conclusions can be drawn about the influence of DBER on instruction
from evaluations of these programs. However, the evaluations do offer
some insights into the broader challenge of translating research into practice, and of accurately measuring faculty members' instructional practices.

Echoing the national survey findings from physics (Henderson and
Dancy, 2009), evaluations from these programs suggest that some have
been more successful in increasing faculty awareness about research-based
practices than in driving actual change in teaching practice. For example,
in an evaluation of the New Faculty Workshop in Physics and Astronomy (Henderson, 2008), 192 former participants said that they used the
DBER-based instructional strategy of peer instruction (Mazur, 1997)—a
collaborative learning approach in which lectures are interspersed with
conceptual questions designed to expose common difficulties in understanding the material. However, only 19 percent of the 192 participants reported
instructional activities that *could* be consistent with the basic features of
peer instruction, as many of them omitted the peer-to-peer interaction
component. These responses suggest that the workshop participants were

---

[6]For more information, see http://pogil.org/ [accessed March 31, 2012].

unaware of the research on the social nature of learning—a key underlying principle of peer instruction. As previously discussed, lack of awareness about the underlying principles of an innovation can lead faculty members to discontinue using that innovation (Rogers, 2003).

Surveys of participants in the National Academies Summer Institute workshops conducted before, shortly after, one year after, and two years after their participation, indicated substantial increases in scientific teaching practices over time (Pfund et al., 2009). However, another investigation of the Summer Institutes and the NSF-funded Faculty Institutes for Reforming Science Teaching Program yielded more mixed results (Ebert-May et al., 2011). That evaluation included surveys and observations of videotaped classes within 6 months of the workshop and again up to 18 months later. On the surveys, more than 75 percent of participants reported frequent use of learner-centered and cooperative learning activities following their workshop training. Yet, analysis of the videotapes using the Reformed Teaching Observation Protocol (Sawada et al., 2002) revealed that the bulk of instruction included pure lecture or lecture with some demonstration and minor student participation (Ebert-May et al., 2011). From the first to the final videotaped class, 25 percent of instructors moved toward more learner-centered practices and 15 percent moved toward more instructor-centered practices. The practices of the remaining 60 percent did not change. A caveat to the findings of Ebert-May et al. (2011) is that alumni of the Summer Institutes frequently commented in surveys that it took three or more years of experimenting with learner-centered teaching strategies before they could implement those strategies effectively (Pfund et al., 2009). These results suggest that measuring the influence of DBER and related research on teaching requires a nuanced, longitudinal model of individual behavior rather than a traditional "cause and effect" model using a workshop or other delivery mechanism as the intervention.

An evaluation of POGIL workshops used a modified version of Rogers' (2003) innovation decision model (see Box 8-1) to identify six stages of readiness to adopt POGIL (Bunce, Havanki, and VandenPlas, 2008). At the start of the workshop, most survey respondents (56 percent) reported that they had already implemented POGIL, while another 30 percent reported plans to adopt this innovation (stages 4 and 5 of readiness to implement). Comparing responses about adoption readiness from pre-survey to post-survey, the results suggest that workshop participation had a limited effect: Nearly half (47 percent) of participants stayed at the same stage of readiness to adopt POGIL, 29 percent increased by one or two stages, and 26 percent decreased by one or two stages. The evaluators interpreted the movement to lower levels of adoption readiness as an indication that, for some participants, the workshops served as a reality check, causing them to more accurately describe their adoption (or nonadoption) of the innovation in the postsurvey (Bunce, Havanki, and VandenPlas, 2008).

In contrast to physics, biology, and chemistry, evaluations of professional development programs in engineering and the geosciences do reveal changes in practice. However, as with other research based on faculty self-reports, these findings must be interpreted with caution. In a survey of National Effective Teaching Institute in Engineering Education alumni who participated between 1993 and 2006, participants credited the workshop with raising their awareness and use of various learner-centered strategies (Felder and Brent, 2010). In the geosciences, a survey comparing faculty who had participated in On the Cutting Edge (either by using the website, or by attending a workshop and using the website), to faculty who had not participated in the program revealed several differences between the two groups. Participants were more likely than nonparticipants to report adding group work or small group activities to their teaching (40 percent of participants compared with 15 percent of nonparticipants); spending less time lecturing (43 percent of participants, compared with 22 percent of nonparticipants) and more time using in-class questioning, small group discussion and in-class exercises; and making more use of education research. In many cases, impacts were more pronounced for those who attended a workshop and made use of the website. Additional, qualitative data indicated that workshop participants underwent a shift in their teaching philosophy to an approach that was more focused on student-centered learning (McLaughlin et al., 2010).

## Discussion of Evaluation Results

The literature on effective professional development in the sciences and engineering including in K-12 education (Loucks-Horsley et al., 2009; Wilson, 2011), along with a review of 191 articles published in peer-reviewed journals (Henderson, Beach, and Finkelstein, 2011) can help to explain the findings from the evaluations of disciplinary professional development programs. This research suggests that successful efforts to translate research into practice include more than one of the following components:

- **Sustained, focused efforts, lasting 4 weeks, one semester, or longer.** This finding implies that one-time workshops or a collection of unrelated workshops are unlikely to be successful (Cohen and Hill, 2001; Garet et al., 2001).
- **Feedback on instructional practice.** Faculty members are more likely to make significant changes in their teaching practice if they receive coaching and feedback when trying a new instructional practice (Henderson, Beach, and Finkelstein, 2011).

- **A deliberate focus on changing faculty conceptions about teaching and learning.** Just as research has shown that deep, conceptual change is often required for students to replace their alternative conceptions with scientifically correct understandings of phenomena (National Research Council, 2007), the research on change in undergraduate science and engineering education shows that faculty are more likely to change their teaching practice when they engage in deep conceptual change (e.g., Gibbs and Coffey, 2004; Ho, Watkins, and Kelly, 2001).

Although the research suggests that a sustained approach enhances effectiveness, some of the professional development programs described above consist of one-time workshops. Recognizing the importance of a sustained approach, the design of the Physics New Faculty Workshop initially included two types of follow-up activities for participants. However, because few workshop participants actually attended these reunions, most of the funding has been reprogrammed to support the single annual workshop (Krane, 2008). The lack of follow-up might explain why this workshop appears more effective in raising awareness of the research than in leading to change in teaching practice.

In contrast, the National Effective Teaching Institute also relies on one-time workshops, yet the evaluations indicated that the workshops did improve instruction. The workshop's evaluators drew on the adult learning literature to explain that workshop's success (Felder and Brent, 2010). Specifically, the workshops meet five important criteria that motivate adult learners (Wlodkowski, 1999):

1. Expert presenters
2. Relevant content
3. Different options for applying recommended methods
4. Praxis (action plus reflection)
5. Group work

Evaluations of On the Cutting Edge, which does not rely on one-time workshops, also indicated that the workshop has led to a moderate level of change in teaching practice. These results might be explained by some key aspects of the program that are consistent with the research on effective professional development (Macdonald et al., 2004). First, the program promotes a sustained approach through its interactive website, which is designed to build ongoing collegial networks and provide ready access to integrated resources that link science, pedagogy, assessments and research on learning (Manduca, Mogk, and Stillings, 2004). Second, the underlying change strategy goes beyond promoting a particular type of finished curriculum or pedagogy. Rather, it engages faculty in reflecting on, and contributing to, curriculum and teaching strategies—primarily through the website.

Despite the challenges of using professional development to translate research into practice, these findings, albeit largely from faculty self-report data, illustrate that it is possible to increase awareness of innovations. A limited amount of evidence suggests that practices can change, although the long-term nature of those changes is not well understood. Moreover, efforts to scale professional development to the level that would influence large numbers of faculty across different institutions are in their early stages.

## Professional Development for Future Faculty in Science and Engineering

Various efforts also are under way to shift the socialization of prospective faculty toward greater commitment to good teaching, including the use of research-based practices (Austin, 2011). This work is based on previous research suggesting that early socialization of future faculty members carries long-term consequences in their professional behavior (Bess, 1978; Clark and Corcoran, 1986).

At the national level, the NSF-supported Center for the Integration of Research, Teaching, and Learning (CIRTL) involves a core network of 6 universities and an expanded network of more than 30 universities to provide professional development opportunities for doctoral students and postdoctoral scholars in science, technology, engineering, and mathematics. Evaluations of the impact of CIRTL-related professional development rely largely on faculty self-report data. These evaluations suggest that participants develop a greater sense of the value of teaching as part of their careers; a wider range of approaches to analyzing teaching problems; and an enhanced ability to encourage student learning (Austin, Connolly, and Colbeck, 2008). Furthermore, participants often indicate that they feel better prepared for undergraduate teaching, have a greater sense of self-efficacy about teaching, and value opportunities to interact with others with similar interests regarding teaching through the learning communities (Austin, 2011). In a similar vein, alumni of the Preparing Future Faculty Program report that the program legitimizes teaching and offers ongoing support through a community of teachers (DeNeef, 2002). Launched in 1993 and sponsored by the American Association of Colleges and Universities, this program prepares doctoral students in various disciplines (including the sciences) for teaching, academic citizenship, and research. No discipline-specific evaluations of the program exist; the overall evaluation includes a survey of 271 alumni (129 responded) and follow-up interviews with 25 survey respondents, so the findings are not conclusive.

Although the evidence from these efforts is too limited to draw conclusions, altering the preparation and expectations of doctoral students for teaching in science and engineering potentially represents a more efficient way to influence future instructional practice than changing the teaching behavior of already active faculty (Fairweather, 2005).

## PUTTING REFORM EFFORTS INTO CONTEXT

Regardless of the availability of quality research and quality professional development to translate that research into practice, change in the teaching practices of science and engineering faculty does not come easily. Teaching behavior—like human behavior generally—is influenced by the contexts in which it is situated (Bronfenbrenner, 1979). Faculty members' teaching decisions depend on the interplay of individual beliefs and values, which have been shaped by their previous education and training, and the norms and values of the contexts in which they work. These contexts include the department, the institution, and external forces beyond the institution (Austin, 2011; Quinn-Patton, 2010; Seymour, 2001; Tobias, 1992).

### Institutional Factors Influencing the
### Translation of DBER into Practice

One possible explanation for the continuing gap between awareness of DBER and adoption of new research-based teaching practices is that the initiatives described in the previous section, led by national organizations, were not sufficiently attuned to factors within academic institutions. Efforts to change science and engineering faculty members' teaching may encounter "local" barriers, such as institutional leadership, departmental peers, reward systems, students' attitudes, and, of course, the beliefs and values of the individual faculty members themselves (Austin, 2011; Fairweather, 2008). These factors have been analyzed extensively in the higher education research literature (see Eckel and Kezar, 2003; Fairweather, 2005; Fisher, Fairweather, and Amey, 2003; Kezar, 2009; Schuster and Finklelstein, 2006). We briefly discuss them here to provide context for efforts to translate DBER into practice.

### The Department

As the immediate context in which faculty members work, the academic department or program has the greatest influence on how faculty members allocate their work time and the decisions they make about teaching (Austin, 1994, 1996). Given the importance of the department, improving learning in undergraduate science and engineering courses may depend as much on research into departmental culture, curriculum content, sequencing, and assignment of teachers to courses as it does on research on the impact of various teaching methods (Fairweather, 2008).

Among other important departmental decisions, class size and physical space can influence the extent to which faculty apply findings from DBER and related research. Research on innovation in physics teaching indicates that large class sizes and the traditional classroom space can pose barriers to faculty adoption of innovative teaching approaches (Henderson and Dancy, 2007). Similarly, geoscience faculty are less likely to use research-based, interactive techniques in large classes than in smaller ones (Macdonald et al., 2005).

In response to these challenges, a few departments and institutions have remodeled classroom space and facilities (see Box 6-1 for some examples).[7] For example, in 1993, Rensselaer Polytechnic Institute applied findings from physics education research to establish a studio physics course in a special classroom designed to support small-group collaboration and problem solving (Cummings, 2008). The redesign was so well-received by students and faculty that, by 2008, all introductory physics courses at Rensselaer (15 to 20 sections, each with approximately 50 students) were studio courses. Early studies showed little improvement in student conceptual understanding or problem-solving skills. Later implementations, which added research-based curricula, resulted in content learning gains over traditional courses (Cummings et al., 1999; Sorensen et al., 2006). To offset the considerable expense of studio courses, the NSF-sponsored Student Centered Active Learning for Undergraduate Programs project (SCALE-UP) supports institutions in restructuring large-enrollment classes following the studio model. By 2011, nearly 100 colleges and universities (or about 2 percent of the 4,400 degree-granting 2- and 4-year institutions in the United States, as counted by the National Center for Education Statistics) had adopted the SCALE-UP approach, with specially designed classrooms for physics, chemistry, mathematics, engineering, and literature courses. As discussed in Chapter 6, learning gains have been shown in many of these courses, especially for female students (Beichner, 2008).

While posing some barriers, the departmental context and culture also may facilitate the productive adoption of DBER findings. For example, faculty members may be more inclined to use DBER in their teaching if they learn about it from disciplinary colleagues, rather than from education researchers (Wieman, Perkins, and Gilbert, 2010). Guided by this assumption, two initiatives at the University of Colorado and the University

---

[7]Project Kaleidoscope is also spearheading an initiative to change learning spaces. Evaluations of those efforts have not yet been published. For more information, see http://www.pkal. org/activities/PKALLearningSpacesCollaboratory.cfm [accessed March 31, 2012].

of British Columbia focus on the department as the key unit of change. These initiatives make significant amounts of funding available to academic departments through a competitive process designed to encourage departmental colleagues to discuss their shared educational goals and to promote the idea that courses belong to the department as a whole (Wieman, Perkins, and Gilbert, 2010). In a 2010 faculty survey, most respondents at the University of Colorado reported undertaking activities that were consistent with the goals of the initiative. Sixty-two percent of respondents reported they had developed learning goals and used those goals to guide instruction, 56 percent reported they used information on student thinking and/or attitudes to guide their teaching, and 47 percent used pre/post measures of learning to inform their teaching practice. In addition, respondents indicated that more than 55 courses incorporated more research-based teaching practices than in the past (Wieman, Perkins, and Gilbert, 2010).

As noted earlier, such self-report data must be interpreted with caution. In addition, these initiatives include some features that might be difficult to replicate, and those features might explain some of the positive results. Specifically, the departments that won funds used most of their money to employ science education specialists, postdoctoral scholars with a doctorate in the discipline who receive intensive training in education and cognitive science. These specialists assist faculty in formulating learning objectives, designing assessments, and creating research-based classroom activities— tasks essential to instructional reform for which research-active faculty may not have time. It is unclear how the reforms will continue as the funds for these postdoctoral positions come to an end.

## Institutional Priorities and Reward Systems

The reward system influences instructional decisions because it sends strong signals about the relative priority of teaching and research to faculty members (for detailed analyses of institutional priorities, time allocations, and salaries see Braxton, Lucky, and Helland, 2002; Fairweather, 2005; Leslie, 2002; Schuster and Finkelstein, 2006). In turn, faculty members respond to these signals. They may be less interested in change efforts related to their teaching if they expect that such efforts will give them less time to do research, and if they perceive research as more highly valued by the institution (Fairweather, 2005, 2008). At minimum, if faculty members are to consider investing time in implementing new pedagogies, they must not feel that such time will be a negative factor in salary and advancement considerations.

Institutional priorities and reward systems influence workloads, including how much time faculty members have to understand and incorporate new teaching strategies based on DBER and related research. Much DBER investigates the learning process itself and identifies the relationships between instruction and student learning. DBER also attempts to uncover what works. However, this basic research is not always (or even often) translated into methods that faculty members can adopt to incorporate the research findings into their instructional practice. Science and engineering faculty members work an average of 55 to 60 hours per week (Fairweather, 2005). As scientists, they may be interested in what DBER tells them about effective instructional practices and about how students learn. Indeed, as we have discussed in this chapter, faculty awareness of research on effective teaching is increasing (Henderson and Dancy, 2009). Yet with their myriad obligations, most faculty members cannot afford to spend an unspecified (and perhaps open-ended) allocation of their work time figuring out how to translate DBER findings into effective and efficient ways to change their instructional practices (Fairweather, 2008). Despite the efforts of university-level professional development offices and professional development efforts like the ones described above, the translation process often remains elusive.

## Students

Because students can exert a strong influence on the learning process (Brower and Inkeleas, 2010; Smith et al., 2004), student responses to new teaching practices informed by DBER and related research may facilitate or discourage adoption of such teaching practices. Students in classes that have been restructured face the difficult and sometimes frustrating task of learning new ways of thinking and solving problems in science or engineering. When participating in more challenging and effective learning experiences, they sometimes complain because they have grown comfortable with being told facts to memorize (Silverthorn, 2006). Cummings (2008) has reported that the students with the strongest academic records are sometimes the most resistant to change, perhaps because they have grown accustomed to earning good grades through memorization and more traditional approaches. Faculty members at institutions where student course evaluations play a role in assessment of their teaching may be reluctant to try new, research-based teaching approaches if they expect that those approaches will lead to critical evaluations.

However, there are documented examples of student course evaluations improving after the adoption of research-based teaching practices (e.g., Hativa, 1995; Silverthorn, 2006). With increased exposure to research-based learning environments, students can become enthusiastic. At North Carolina State University, hundreds of students have taken physics classes that have been restructured from lectures and laboratories into the SCALE-UP classrooms described above. Students who have taken a first-semester SCALE-UP class universally select the SCALE-UP version, rather than the lecture version, for their second semester physics course. Students report that their friends direct them into SCALE-UP classes, and SCALE-UP sections fill before the lecture sections. In focus groups of students who had taken the lecture version for their first semester and SCALE-UP in their second semester, students reported that they were learning at a deeper conceptual level in the SCALE-UP class; these perceptions were supported by evidence of gains in learning and performance (Beichner, 2008).

### Faculty Members' Beliefs

Individual faculty members have strongly held beliefs and conceptions of teaching in general and of their own teaching in particular that may influence the extent to which they adopt research-based practices (see Blackburn and Lawrence, 1995; Leslie, 2002; and Schuster and Finkelstein, 2006, for reviews of the role of faculty members' beliefs). Indeed, one review of the literature on change strategies in science and engineering found that faculty beliefs were among the most common barriers to developing reflective teachers (Henderson, Beach, and Finkelstein, 2011).

More specifically, in one interview study of 30 physics faculty, the faculty members identified learning goals for their students that largely aligned with the research on effective teaching and learning (Yerushalmi et al., 2010). Also consistent with the research, they identified problem features that would support those learning goals. Nevertheless, many of the faculty members reported that they did not use the problem features they identified, because the features conflicted with the instructors' strongly held values about minimizing student stress and presenting problems clearly during exams. In contrast, Gess-Newsome et al. (2003) observed that science faculty who experienced a mismatch between their personal beliefs and their teaching practices were more likely to be dissatisfied with their teaching and more open to developing new knowledge and beliefs. These mixed findings indicate that role of faculty members' beliefs is ripe for further exploration.

## The Change Process in Undergraduate Science and Engineering

Broader research on change strategies in higher education also can shed light on efforts to translate DBER and related research into practice. One review identified three communities involved in efforts to promote change in undergraduate science and engineering education (Henderson, Beach, and Finkelstein, 2011) as follows:

1. DBER scholars (referred to as the science education research community). Scholars in this community typically examine teaching and learning within the disciplines, and are situated in disciplinary departments.

2. The faculty development research community. Scholars in this community conduct, evaluate, and improve faculty professional development. These scholars are often located at campus centers for teaching and learning.

3. The higher education research community. Scholars in this community often investigate institutional and organizational culture, climate, and policies, and are found within a school or college of education.

Each of these communities operates relatively independently, rarely communicating with the other communities or trying to build on their research methods or findings (Henderson, Beach, and Finkelstein, 2011). Indeed, as Figure 8-1 illustrates, although the change efforts described in the literature shared the common goal of changing undergraduate teaching in science and engineering, they reflected different change strategies. DBER papers emphasized the strategy of disseminating curriculum and pedagogy, faculty development research sought to develop reflective teachers, and

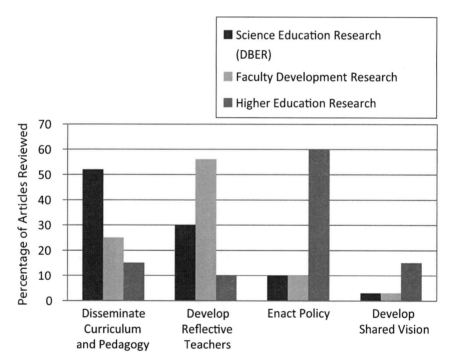

FIGURE 8-1 Change strategies used by different research communities (as summarized by Henderson, Beach, and Finkelstein, 2011).

higher education research was largely concerned with policy. This distribution led Henderson, Beach, and Finkelstein to observe that the three communities would benefit from greater communication to share research methods and findings.

In addition to fostering greater communication among the research communities seeking to influence instructional practices in undergraduate science and engineering, these efforts must attend to the broader system in which that education takes place (Austin, 2011; Quinn-Patton, 2010). A systems approach to change in higher education shows that multiple factors influence faculty members' choices about their teaching practice (see "Factors Influencing the Translation of DBER into Practice" in this chapter). While some factors may encourage use of new approaches to teaching, other factors may simultaneously discourage such innovative practice. Thus, the multiple contexts within which faculty members work, and the influences and interactions of various features of higher education institutions and systems, must be considered to more fully understand the translation of DBER and related research into practice.

Approaches to change that take a linear path, relying on one factor or intervention alone, are unlikely to yield the desired outcome. Given the complexity of higher education institutions, change efforts are most effective when they use both a "top-down" and a "bottom-up" approach, take into consideration the relevant factors that affect faculty work, and strategically use multiple change levers (see Austin, 2011, for a summary of this literature).

## SUMMARY OF KEY FINDINGS

- *Faculty instructional practices in science and engineering have not been systematically documented. National surveys exist for faculty in general and separately in the geosciences, engineering, and physics. The results of the discipline-specific surveys are not always nationally representative and the conclusions are based on self-report data. These surveys suggest that lecture is the primary mode of instruction in the sciences.*
- *Available research on efforts to translate DBER and related research into practice mostly consists of evaluations of professional development programs in physics, biology, the geosciences, chemistry, and engineering, which largely rely on self-report data. For the most part, these evaluations suggest that translational efforts have been more effective at raising awareness of research-based practices than at changing practice.*
- *Professional development initiatives that have led to (self-reported) changes in practice are sustained over time, are consistent with*

*research on motivating adult learners, engage faculty in reflection of their teaching practices, and include a deliberate focus on changing faculty conceptions about teaching and learning.*

- *Multiple factors interact to affect faculty instructional decisions. Efforts to translate DBER into practice should take into account individual, departmental, and institutional influences on instruction.*

- *Although the evidence on efforts to prepare future faculty to incorporate research-based practices into their instruction is not conclusive, such efforts could represent a significant leverage point to translate DBER and related research into practice.*

## DIRECTIONS FOR FUTURE RESEARCH

Although it is inherently difficult and complex to assess the extent to which DBER has informed instruction, the understanding of this topic is particularly limited because the research base is particularly sparse. The first step is to develop and test a model of what influences the teaching practices of individual science and engineering faculty members and of how teaching is situated in the larger organizational context. Because both individual and contextual aspects are relevant to assessing the effects of DBER, both are relevant to increasing the effects of DBER on teaching in the future.

A reliable baseline understanding of faculty instructional practices in the sciences and engineering also is needed; this research should mitigate the limitations of self-report data to the extent possible and provide insights into variations by discipline, institutional type, and course type (e.g., large courses versus small, introductory courses, courses for majors). In a related vein, current research on discipline-specific professional development efforts to translate DBER and related research into practice mostly consists of program evaluations. By their nature, these evaluations do not rigorously examine the question of how such programs influence instruction across disciplines. Future research on this topic should take into account the larger body of research on adult learning and effective professional development to identify guiding principles for the effective translation of research into practice. Specifically, it would be productive to study what support, guidance, and knowledge of underlying principles faculty need to successfully implement research-based practices (Henderson and Dancy, 2009).

Because departmental and institutional norms and cultures reflect shared values and beliefs—including beliefs about teaching and learning—gaining understanding of departmental and institutional norms and cultures is an essential step in designing efforts to enact new policies supporting educational innovation. Based on this understanding, policy initiatives can be aligned with existing norms and/or be carefully tailored to modify those shared norms that pose barriers to the use of DBER and related research. And finally, among the 191 studies reviewed by Henderson, Beach, and Finkelstein (2011), none examined initiatives that included significant changes to faculty recognition and reward systems. Considering that many change initiatives are slowed by barriers within institutional recognition and reward systems, it is important to develop approaches to overcome these barriers. This gap in the research on change initiatives that are based on DBER and related research represents an important opportunity for future study.

# 9

# Future Directions for Discipline-Based Education Research: Conclusions and Recommendations

The United States currently faces a great imperative to improve science and engineering education. U.S. colleges and universities play a vital role in preparing a diverse technical workforce and a science-literate citizenry, and they must provide sustained attention to motivating, engaging and supporting the learning of all students who enter college science and engineering classrooms. Meeting this imperative requires a deep understanding of how people think, learn, and feel about natural and physical processes and phenomena. Discipline-based education research (DBER), which combines the expertise of scientists and engineers with methods and theories that explain learning, helps to provide this understanding. The DBER enterprise already has generated insights into how students learn in a discipline and into effective instructional strategies that can prepare more students to address current and future societal challenges.

DBER investigates learning and teaching in a discipline using a range of methods with grounding in the discipline's priorities, worldview, knowledge, and practices. It is informed by and complementary to research on human learning and cognition. The long-term goals of DBER are to

- understand how people learn the concepts, practices, and ways of thinking of science and engineering;
- understand the nature and development of expertise in a discipline;
- help to identify and measure the efficacy of appropriate learning objectives and instructional approaches that advance students toward those objectives;

- contribute to the knowledge base in a way that can guide the translation of DBER findings to classroom practice; and
- identify approaches to make science and engineering education broad and inclusive.

DBER can be a field of study within any academic discipline, in the sciences and beyond. However, because this study focused on education research in a select set of science and engineering disciplines—physics, chemistry, engineering, biological sciences, the geosciences, and astronomy—this report uses the term DBER to refer only to these disciplines.

The previous chapters have described the current status of DBER; synthesized peer-reviewed, empirical research on undergraduate teaching and learning in the sciences and engineering; and examined the extent to which this research currently influences undergraduate science and engineering instruction. By presenting conclusions and recommendations that draw on the key findings and directions for future research from previous chapters, we describe here the intellectual and material resources that are required to further develop DBER. The conclusions are grouped into four areas:

1. Defining DBER
2. Synthesizing DBER
3. Translating DBER findings into practice
4. Advancing DBER as a field of inquiry

We end the chapter with recommendations to enhance the impact of the findings from DBER and advance the fields of DBER, and by proposing a future research agenda for DBER.

## DEFINING AND DESCRIBING DISCIPLINE-BASED EDUCATION RESEARCH

*Conclusion 1: At present, DBER is a collection of related research fields rather than a single, unified field. Most efforts to develop and advance DBER are taking place at the level of the individual fields of DBER.*

The term DBER is best thought of as an overarching term that refers to a set of distinct fields that have emerged over several decades across multiple disciplines. The individual fields of DBER share the overall goal of improving learning and teaching in a discipline through the use of findings from empirical research. To meet this goal, researchers in the different fields of DBER build on some common theoretical approaches to learning and often employ similar research methods. However, each field is tightly coupled to its parent discipline, which gives rise to differences across the fields of DBER, such as in the history of development, professional pathways for researchers, and emphasis of research.

As described in Chapter 2, the fields of DBER share some common milestones in their development that reflect the larger context of science and education. Yet the developmental trajectories of the DBER fields differ. Physics education research was established earliest, followed by chemistry education research and then engineering education research. Biology, the geosciences, and astronomy education research have emerged more recently. The fields that emerged later appear to have benefitted from building on and borrowing from the more established fields in DBER, especially from physics education research.

The parent disciplines for each field differ in terms of how readily they have embraced research in education, and the availability of venues for publishing research. Chapter 2 describes how such differences continue to shape the way research is conducted in each field of DBER and the paths that scholars can follow to gain expertise in DBER.

*Conclusion 2: The fields of DBER have made notable progress in establishing venues for publishing and in gaining recognition from their parent disciplines. However DBER scholars still face challenges in identifying pathways for training and professional recognition.*

Each DBER field has one or more professional organizations that support education research through policy statements, publication venues, and conferences. As discussed in Chapter 2, many of these professional homes are sections of larger disciplinary professional societies.

The number of journals that publish DBER varies by field, but currently all of the fields have at least one peer-reviewed journal that publishes DBER. Some tension exists between publication venues intended to share research findings among researchers and venues intended to inform instructors of the findings of DBER that might be useful in their classrooms.

Pathways to establish interdisciplinary research such as DBER are not straightforward. Tenure and promotion committees may not take into account the time and energy necessary to become acculturated into a new field, which poses particular challenges for nontenured DBER faculty. In a different vein, institutions and disciplinary departments do not always recognize the distinction between education specialists whose primary focus is on teaching and DBER scholars who conduct research on teaching and learning. As a result, expectations for DBER faculty regarding teaching, research, and service can sometimes be imbalanced (see Chapter 2).

*Conclusion 3: DBER encompasses a range of research goals and emphases that span the continuum from basic research that provides insights into fundamental learning processes to applied research on effective designs for instruction carried out in actual classrooms.*

Although an overarching goal of DBER is to improve undergraduate learning and teaching, individual studies do not always have a directly applied component. Instead, as in other areas of science, the DBER that is synthesized in Chapters 4 through 7 includes a blend of studies with immediate application in the classroom and those that explore more basic questions. Basic research in DBER might examine why students hold particular understandings, beliefs, and ideas (some of them incorrect) or why one instructional intervention is more effective than another.

*Conclusion 4: High-quality DBER combines expert knowledge of a science or engineering discipline, of learning and teaching in that discipline, and of the science of learning and teaching more generally.*

A long-term goal of DBER is to understand how people learn science and engineering in order to improve learning and teaching. Research that advances this goal must be grounded in an understanding of what it means to develop expertise in a discipline and the challenges inherent in developing that expertise. At the same time, individual studies must be informed by a working knowledge of existing findings related to learning and teaching in a discipline, and more broadly, an understanding of the methods that are appropriate for investigating human thinking, motivation, and learning. These methods are often drawn not from the parent science or engineering discipline, but from disciplines in the behavioral and social sciences such as psychology, sociology, anthropology, and education.

Bringing together the diverse expertise required poses a challenge that can be met in a variety of ways. As described in Chapter 2, individual researchers can begin to develop the necessary expertise through well-crafted graduate and postdoctoral programs. The required integration of expertise can also be accomplished through collaborations that range from two individuals in different disciplines (e.g., physics and psychology) to larger, strategically assembled teams of researchers.

*Conclusion 5: Conducting DBER and promoting change by applying the findings from DBER to improve instruction are distinct but interdependent pursuits.*

In Chapter 2, the analysis of the DBER fields using Fensham's (2004) criteria for characterizing the emergence of new disciplines reflects the predictable tension between advancing the research itself and increasing the use of DBER findings. Many DBER scholars, their disciplinary colleagues, professional societies, and funding agencies are motivated by the need to reform science and engineering education in ways that are informed by DBER findings. And, as in any discipline, DBER scholars strive for high-quality research. Education research centers, funding programs, and some journals blend both of these goals. Clearly articulating the distinction between discipline-based education research and the application of DBER findings—and embracing the value of both—is important for ensuring continued advancement of the research, promoting improvement in under-graduate education, and enhancing synergies between these efforts.

## SYNTHESIZING DISCIPLINE-BASED
## EDUCATION RESEARCH

*Conclusion 6: In all disciplines, undergraduate students have incorrect ideas and beliefs about fundamental concepts. Students have particular difficulties with concepts that involve very large or very small temporal or spatial scales, in part because they lack an experiential basis from which to develop an understanding about these concepts. Not all incor-rect ideas and beliefs are equally important in terms of understanding students' learning in a discipline, however. Across all disciplines, the education community needs a better understanding of those that pose the biggest challenges to learning at the undergraduate level, how they arise, and how to help students align them with scientific explanations.*

Undergraduate science and engineering learning, like all learning, occurs against the backdrop of prior knowledge. Students at all levels, from preschool through college, enter instruction with various commonsense, but incorrect, interpretations of scientific and engineering concepts, as well as personal beliefs that can affect their learning.

Similar to researchers at the K-12 level, DBER scholars have devoted considerable time to investigating undergraduate students' conceptual understanding. DBER studies have identified a wide range of incorrect ideas and beliefs related to such fundamental concepts as electricity, magnetism, the nature of matter, phase changes, evolution, and deep time. As discussed in Chapter 4, DBER clearly documents students' difficulties in understand-ing interactions that involve very large or very small spatial or temporal scales. Notable examples include misunderstandings of Earth's history and myriad learning challenges in chemistry that result from difficulties in understanding that matter is made of discrete particles.

The most productive lines of research involve concepts that are central to the discipline and focus on incorrect understandings that are widely held and resistant to change. By drawing on expert knowledge of which concepts are central to a discipline and expert understanding of those concepts, DBER offers a unique contribution to this research. DBER scholars in engineering, biology, the geosciences, and astronomy are beginning to identify incorrect ideas and beliefs to determine which concepts are more difficult to learn than others. This research often is coupled with instructional techniques that are targeted at eliminating a specific erroneous belief. In physics and chemistry, some DBER scholars have begun exploring whether some classes of misconceptions are connected by a common underlying cognitive structure. Identifying such common structures may facilitate the development of instructional strategies that address large classes of misconceptions, rather than addressing them one at a time.

Physics education research has shown that several types of instructional strategies can promote conceptual change, or help to align students' understandings with scientific explanations. These strategies, described in Chapter 4, include interactive lecture demonstrations, interventions that target specific erroneous beliefs or incorrect ideas, and introduction of linking concepts to bridge students' incorrect idea with the accepted scientific explanation.

> *Conclusion 7: As novices in a domain, students are challenged by important aspects of the domain that can seem easy or obvious to experts, such as complex problem solving and domain-specific representations like graphs, models, and simulations. These challenges pose serious impediments to learning in science and engineering, especially if instructors are not aware of them.*

The ability to solve complex problems is central to science and engineering. Problem solving has been extensively studied in cognitive science, physics education research, chemistry education research, and engineering education research. It is an emerging area of study in biology education research and geoscience education research. A considerable amount of cognitive science and discipline-based education research addresses well-defined quantitative problems. Except in engineering and chemistry, considerably less research exists on ill-defined, open-ended, or context-rich problems, which are more characteristic of what scientists and engineers encounter in their professional lives. Chapter 5 shows that across the disciplines, students have difficulty with all aspects of problem solving and they approach problem solving differently than experts.

Equations, graphical displays, diagrams, and other representations feature prominently in problem solving and other scientific and engineering activities. The disciplines differ in terms of how problems are specified and the conventions for representation, and many of these representations are unique to a given discipline (e.g., Lewis structures in chemistry, cladograms in biology, and models of the Earth's structure in the geosciences). To flourish in science and engineering courses and careers, students must become fluent with the discipline-specific approaches and representations used by experts in the field. Students begin this process as novices, and, with targeted assistance, can move toward expert-like understanding. Along the way, how students create, use, and interpret representations can provide insight into their understanding of important concepts in a discipline.

Although equations, graphical displays, and other representations may seem easy to understand for undergraduate faculty who are domain experts, the research discussed in Chapter 5 shows that students have difficulty extracting information from these representations and constructing appropriate representations from existing information, regardless of discipline. For example, in chemistry, students have difficulty constructing particulate-level diagrams of chemical and physical phenomena. Students also have difficulty understanding the commonality of the underlying structure across different representations of the same phenomenon, such as imagining the three-dimensional distribution of earthquake epicenters when given a map showing earthquake depth and magnitude using both colors and symbols.

*Conclusion 8: Improving undergraduate science and engineering education involves integrating proven strategies for general instruction with strategies designed to explicitly target challenges that are unique to science and engineering or to a specific discipline.*

As discussed in Chapter 6, a considerable amount of DBER examines instruction that is based on established learning theories and principles. Consistent with research from cognitive science, educational psychology, and science education, DBER indicates that involving students actively in the learning process can enhance learning more effectively than traditional instructional methods, such as lecturing by a professor. Exemplary methods include making lectures more interactive, having students work in groups, incorporating authentic problems and activities, and promoting metacognition. These strategies are not discipline-specific, and range in scope and complexity from slight modifications of instructional practice—such as beginning a lecture with a challenging question for students to keep in mind—to completely redesigning the learning space. Overall, DBER does not yet provide evidence about the relative effectiveness of various student-centered strategies or whether any of these strategies are differentially effective for learning certain types of content. The findings and the gaps in current understanding discussed in Chapter 6 suggest that effective instruction includes a range of well-implemented, research-based approaches.

Discipline-specific research in physics and chemistry indicates that students can be taught more expert-like problem-solving skills and that scaffolding, or providing appropriately structured support to learners, appears to be beneficial in this regard. Similarly, Chapter 5 discusses specific instructional strategies that have been shown to help students create, use, and interpret graphical representations. In all disciplines, instructors can improve students' learning by building some of these approaches into their teaching.

*Conclusion 9: The use of learning technology in itself does not improve learning outcomes. Rather, how technology is used matters more.*

Chapter 7 shows that evidence on the efficacy of widely used technologies such as animations and personal response systems (clickers) is mixed. Clickers are small handheld devices that allow students to send information (typically their response to a multiple choice question provided by the instructor) to a receiver, which tabulates the classroom results and displays the information to the instructor. The most compelling evidence on their use shows that learning gains are associated only with applications that challenge students conceptually and incorporate socially mediated learning techniques, such as having students work and be assessed collaboratively. The use of animations also has been studied and shown to enhance learning in some circumstances, but to be ineffective or even detrimental to students' learning in other situations. Taken together, this research demonstrates that how technology is used matters more than simply using technology. For technology to be effective, instructors must be aware of the conditions that support the effective use of technology and incorporate it into their lessons with clear learning goals in mind.

*Conclusion 10: Across all disciplines and all topics of inquiry in DBER, relatively few studies explore whether or how learning and responses to different instructional approaches vary by key characteristics of students such as gender, ethnicity, and socioeconomic status. As a result, current knowledge of similarities and differences among student populations is severely limited.*

With few exceptions, DBER has not examined variation across different populations of students, such as those with different demographic characteristics or ability levels. Similarly, very little DBER conducted in the context of introductory courses distinguishes among outcomes for majors versus nonmajors.

The relative lack of attention to group or individual differences reflects, in part, the foci of DBER to date. Early DBER has studied major trends in learning and teaching before undertaking explorations of subgroups. In addition, in some cases the sample sizes of certain groups are too small to provide statistical power. These gaps preclude a complete and nuanced understanding of undergraduate science and engineering education.

## TRANSLATING DISCIPLINE-BASED EDUCATION RESEARCH FINDINGS INTO INSTRUCTIONAL PRACTICE

*Conclusion 11: Determining the extent to which DBER findings have been translated into instructional practice requires more nuanced, multifaceted investigations than are currently available.*

Determining the extent to which DBER has informed teaching practice is difficult for many reasons. First, as discussed in Chapter 8, a limited empirical baseline exists to document faculty members' instructional practices in science and engineering. Few studies have rigorously examined instructional practices within disciplines, and even fewer have studied practices across disciplines at the undergraduate level. Second, because faculty members may draw on similar findings from DBER, cognitive science, educational psychology, science education, education, and/or the scholarship of teaching and learning to inform their practice, it is difficult to disentangle the effects of DBER from related research fields. Third, DBER and related research can influence teaching practices to varying degrees, from increased awareness of students' learning challenges to complete transformation of instructional approaches. And finally, as research on higher education policy and organization has shown, instructional decisions—including the decision to incorporate DBER and other research—are influenced by many more factors than the mere availability of research findings. The factors discussed in Chapter 8 include institutional leadership, departmental peers, reward systems, students' attitudes, and, of course, the beliefs and values of the individual faculty members themselves. Any study of DBER's translational role must take these challenges and factors into account, and must rely on more than faculty reports of their instructional practices.

*Conclusion 12: DBER and related research have not yet prompted widespread changes in teaching practice among science and engineering faculty. Strategies are needed to more effectively promote the translation of findings from DBER into practice.*

To date, the most common strategy for translating DBER into practice has been to develop new teaching approaches and materials, research them, and then make the most promising ones available to faculty, primarily through workshops. Relying largely on faculty self-report data, evaluations of programs that use this approach in physics, the geosciences, biology, and chemistry indicate that they have generally been more successful in making participants aware of existing research than in convincing participants to adopt new, research-based teaching practices (see Chapter 8). Even for the programs that appear to have influenced practices, the durability of those changes is not well documented. Moreover, efforts to scale professional development to the level that would influence large numbers of faculty across different institutions are still in the early stages.

As discussed in Chapter 8, some initiatives have attempted to shift the socialization of prospective faculty toward greater commitment to good teaching, including the use of research-based practices. Although the evidence from these efforts is still too limited to draw conclusions, altering the preparation and expectations of doctoral students for teaching in science and engineering potentially represents a more efficient way to influence future instructional practice than changing the teaching behavior of already active faculty.

> *Conclusion 13: Efforts to translate DBER and related research into practice are more likely to succeed if they are (1) consistent with research on motivating adult learners, (2) include a deliberate focus on changing faculty conceptions about teaching and learning, (3) recognize the cultural and organizational norms of the department and institution, and (4) work to address those norms that pose barriers to change in teaching practice.*

The research discussed in Chapter 8 suggests that faculty members are unlikely to change their teaching practice without opportunities to reflect on their own teaching practice, compare their practice to research-based, more effective approaches, and become dissatisfied with their own practice. This process of conceptual change for faculty parallels the process of conceptual change to help students develop scientifically correct understandings of natural phenomena.

## ADVANCING DISCIPLINE-BASED EDUCATION RESEARCH AS A FIELD OF INQUIRY

> *Conclusion 14: Ph.D. programs and postdoctoral opportunities in the individual fields of DBER are important mechanisms to provide high-quality education for future DBER scholars.*

Scholars have entered DBER through a variety of pathways, including "border crossing" by researchers in the parent discipline who develop expertise in education research, postdoctoral opportunities, and Ph.D. programs that combine training in the parent discipline with training in education research (see Chapter 2). In the past, the prevalence of these pathways has varied across the fields of DBER. As DBER gains status in the parent disciplines and the fields mature, Ph.D. programs and postdoctoral opportunities are increasingly important to prepare future scholars and advance the research. These two pathways provide mechanisms to systematically integrate knowledge and methodologies from the parent discipline with those from education research, and to enculturate new scholars into a broader professional community.

> *Conclusion 15: Collaborations among the fields of DBER, although relatively limited, have resulted in shared methodology and shared insights into achieving instructional change and building students' understanding of science and engineering. Understandings and perspectives from cognitive science, educational psychology, social psychology, organizational change, education, science education, and psychometrics have similarly enhanced the quality of DBER.*

Collaboration across the fields of DBER has varied by discipline and over time. For those fields that emerged first, particularly physics education research, opportunities for collaboration were limited because so few DBER scholars were active in other disciplines. Research in these early fields did, however, draw on theories and findings from psychology—particularly cognitive psychology, and subsequently, cognitive science. DBER fields that emerged later have benefited from the more established DBER fields by using specific findings and gaining guidance on how to build the field.

Opportunities for interaction across DBER fields are increasing through journals and through meetings and conferences that bring DBER scholars together (see Chapter 2). These kinds of opportunities hold promise for advancing DBER because they enable scholars to share findings and build an understanding of which findings can be applied across disciplines and which are discipline-specific.

As the syntheses in Chapters 4 through 7 illustrate, interaction and collaboration between DBER and related disciplines also are inconsistent. Drawing on research in related disciplines such as cognitive science, psychology, sociology, and K-12 science education is important for several reasons. First, research in these disciplines can provide theoretical frameworks for explaining basic principles of cognition, learning and teaching with which DBER must be consistent. Second, established findings in these disciplines can help to shape and refine the questions that DBER scholars pose, and serve as corroborating evidence for findings in DBER. Third, these disciplines have long traditions of studying human cognition and learning and have developed robust methods for doing so. These methods can prove valuable for DBER scholars.

Collaborations among DBER scholars and researchers in these related disciplines, although relatively limited, have been productive. For example, geoscience and geography education researchers collaborated with psychologists to produce a report on spatial thinking (National Research Council, 2006), and collaboration between geoscience education research and psychology continues through the National Science Foundation-supported Spatial Intelligence and Learning Center. As the fields of DBER advance, these kinds of collaborations as well as collaborations and interactions across fields within DBER merit continued support.

*Conclusion 16: Advancing DBER requires a robust infrastructure for research that includes adequate and sustained funding for research and training, venues for peer-reviewed publication, recognition and support within professional societies, and professional conferences.*

As with any field of research in the sciences and engineering, funding is an essential element of a robust DBER infrastructure. However, as Chapter 2 demonstrates, funding across the fields of DBER is uneven. Adequate support to enable the growth of DBER includes funding for the kinds of studies identified in the research agenda at the end of this chapter, training for Ph.D. students and postdoctoral candidates, and ongoing professional development for active faculty. Continued funding to support programs and initiatives that are designed to translate DBER findings into practice also is important.

The number of venues for publishing empirical research has expanded as the DBER fields have matured. Within a given DBER field and across fields, these journals vary in their standards for research. Because advancing research and applying the findings of this research are important goals of DBER, it is important to strike a balance between journals that publish empirical research primarily to share findings among researchers and journals that publish research in formats accessible to those interested in applying the findings.

Recognition from professional societies that DBER is a viable research field in the science discipline can be important for advancing research and for attracting scholars to the specialty. Such recognition is distinct from acknowledgement that science education in general is important. The fields of DBER all have been recognized by professional societies in the parent discipline as valid and important fields of research.

## RECOMMENDATIONS

Based on our findings and conclusions, we offer a series of recommendations to advance DBER as a field of inquiry, increase the use of DBER findings, and develop a research agenda for DBER. Enacting these recommendations will involve numerous stakeholders, primarily by enhancing their relevant individual efforts, but also by promoting more collaboration among them.

## Advancing Discipline-Based Education Research as a Field of Inquiry

Advancing the individual fields of DBER and DBER as a whole requires simultaneously supporting current DBER scholars and adequately preparing future DBER scholars. These efforts involve providing institutional, material and intellectual support and recognition.

> **Recommendation 1:** In their respective roles, science and engineering departments, professional societies, journal editors, funding agencies, and institutional leaders should clarify expectations for DBER faculty positions, emphasize high-quality DBER work, provide mentoring for new DBER scholars, and support venues for DBER scholars to share their research findings at meetings and in high-quality journals.

## Translating Discipline-Based Education Research into Practice

The committee's recommendations for translating DBER findings into practice involve broader changes to higher education institutions and systems. Implementing these recommendations should blend a top-down and bottom-up approach, take into consideration the factors at work within the multiple contexts that affect faculty members, and strategically use multiple levers for effecting change (see Chapter 8 for a more detailed discussion). Approaches to change that are restricted to a linear path, relying on one factor or intervention alone, are unlikely to lead to the desired outcome.

> **Recommendation 2:** With support from institutions, disciplinary departments, and professional societies, current faculty should adopt evidence-based teaching practices to improve learning outcomes for undergraduate science and engineering students.

> **Recommendation 3:** To increase the future use of DBER-based teaching approaches, institutions, disciplinary departments, and professional societies should work together to prepare future faculty who understand the findings of research on learning and evidence-based teaching strategies, and who value effective teaching as part of their career aspirations.

Recommendation 4: Institutional leaders should include learning and evidence-based teaching strategies in the professional development of early career faculty, and then include teaching effectiveness in evaluation processes and reward systems throughout faculty members' careers. Disciplinary societies and the education research communities within them should support these efforts at the national level.

### Research Agenda for Discipline-Based Education Research

DBER already has added to current understanding of how people learn in the disciplines of science and engineering. Much of this research overlaps with and builds on findings and approaches from cognitive science, K-12 education, and other related fields. Discipline-based education researchers, with support from government and private funding entities, can further enhance the value and impact of DBER by conducting additional research that builds on the directions for future research discussed in Chapters 4 through 7. Specifically, the following cross-cutting themes emerged from those discussions:

- **Research is needed to explore similarities and differences among different groups of students.** With few exceptions, across all disciplines and all topics addressed in this report, there is a dearth of research that explores potential differences among different student populations (e.g., race/ethnicity, gender, majors vs. nonmajors, students of different abilities, etc.).
- **Research is needed in a wider variety of undergraduate course settings.** Existing DBER provides excellent insights into students' understanding and learning of introductory course material. However, gaps remain in the understanding of student learning in upper division courses. In addition, for most of the disciplines in this study, understanding of how students learn in laboratory and field settings is minimal. Activities in which students conduct inquiry on large, professionally collected datasets (such as genomics data and those served by the U.S. Geological Survey, National Oceanic and Atmospheric Administration, National Aeronautics and Space Administration, and various university consortia) have grown in prominence in recent years, but have been understudied. It also is important to augment current understanding of which field activities generate different kinds of learning and which teaching methods are most effective for different audiences, settings, expected learning outcomes, or types of field experiences. DBER scholars also should explore K-12, graduate, and informal education, as appropriate.

- **Longitudinal studies, including those that investigate the transition from K-12 schooling to undergraduate programs, are required to fully understand some phenomena and outcomes that are important to science and engineering education.** Longitudinal studies would enhance current understanding of concepts such as the transfer (or lack thereof) of knowledge from one setting to another, including from K-12 to undergraduate education; the persistence of incorrect ideas and beliefs, and the process of conceptual change; and the effects of student-centered learning on longer-term outcomes such as retention of conceptual knowledge and attitudes about science and engineering. Longitudinal studies also could yield insights into reasons for retention in or departure from science and engineering majors.

- **DBER should measure a wider range of outcomes and should explore relationships among different types of outcomes.** As Chapters 4 through 7 indicate, the vast majority of DBER measures gains in students' knowledge, conceptual understanding, or academic performance. Rigorous research is needed to examine outcomes associated with the affective domain, including students' attitudes about learning. Moreover, it would be helpful to explore relationships among different outcomes, such as the relationship between certain types of skills (e.g., problem solving, spatial ability, competence with science and engineering practices) and outcomes such as students' dispositions toward science and engineering, persistence in the major, or overall understanding of scientific concepts and disciplines. And finally, given the importance of and interest in recruiting and retaining students in the sciences and engineering, additional research is needed that examines outcomes that may provide insight into these issues. Some of these outcomes include declaring a major, decisions to pursue further study in the discipline, and skills in the practices of science and engineering.

- **The emphasis of research on instructional strategies should shift to examine more nuanced aspects of instruction.** The research on instruction described in Chapters 4 through 6 demonstrates that student-centered learning can be more effective than traditional lecture. Now, a more nuanced view of instructional strategies is needed to advance knowledge of student learning in the sciences and engineering. Existing DBER should be expanded to address a broader range of pedagogical techniques and learning progressions that promote conceptual change by moving students toward scientifically normative conceptions; to identify a range of instructional approaches that might help students to use visualizations or solve problems; to describe which kinds of learning environments promote metacognition; and to identify more effective means of engaging students in the practices of science and engineering. In addition research should address whether certain strategies are more or less effective for different types of learners, as well as the learning conditions that appear to support successful outcomes.

- **Better instruments are needed to measure a variety of outcomes.** As discussed in Chapter 3, concept inventories have proliferated across the disciplines in this study. Although they are useful for identifying a suite of previously articulated misunderstandings within a group, they have limitations. To probe student understanding more deeply, faculty and DBER scholars need additional tools for qualitative and quantitative analyses that are widely available and easy to use. As discussed in Chapters 4 through 7, it would be helpful to have instruments for assessing skills that well-designed laboratory instruction can promote, in addition to tools that better measure metacognition, the transfer of knowledge, and gains in spatial thinking and interpretation of representations in the context of undergraduate science and engineering courses. Another pressing need is for instruments that will allow instructors to measure problem-solving skills for large numbers of students in an authentic classroom setting. However, even the best multiple-choice instruments are relatively coarse and can yield inconsistent results (Huffman and Heller, 1995). DBER scholars should recognize the need to continually extend the resolution of these instruments, through such mechanisms as follow-up interviews in which students explain their choices and thinking processes.

Looking across the body of DBER as a whole, the committee also recommends that

- **Additional basic research in DBER is needed on teaching and learning in undergraduate science and engineering.** Most of the studies discussed in Chapters 4 through 7 measure specific outcomes, such as whether particular interventions lead to greater learning gains, or how well students perform on a task or use representations. As those chapters showed, fewer studies focus on "how" or "why" a specific phenomenon or outcome occurs, such as how (if at all) different student populations vary or why technology is not always effective. By asking these questions of "how" and "why," basic research provides the foundation for designing effective learning environments, curricula, or instructional materials, and for making modifications when circumstances change. Decades of basic research in educational psychology and cognitive science have generated theoretical understandings that can be used to examine successes and failures and to guide future research and development efforts in DBER.

- **Interdisciplinary studies are needed to examine cross-cutting concepts and cognitive processes.** DBER scholars have no shortage of discipline-specific problems and challenges to study, but cross-cutting concepts (such as energy or systems) and structural or conceptual similarities that underlie discipline-specific problems (such as concepts in different disciplines that involve very small or very large scales of measurement, or deep time) also merit attention. Interdisciplinary studies could help to increase the coherence of students' learning experience across disciplines by uncovering areas of overlap and gaps in content coverage, and could facilitate an understanding of how to promote the transfer of knowledge from one setting to another.

- **More investigations are needed of teaching and learning across multiple courses in a discipline.** Most of the research that the committee reviewed focused at the level of a single course. Cross-sectional studies of multiple courses within a discipline, or of all courses in a major, would enhance the understanding of how people learn the concepts, practices, and ways of thinking of science and engineering and of the nature and development of expertise in a discipline.

- **Additional research is needed on the translational role of DBER.** To achieve the goal of translating DBER into practice, some research needs to examine organizational and behavioral change. Such studies should draw on existing research on higher education organization and policy examining the influences on faculty decision-making. That research could inform, and enhance the effectiveness of, future efforts to increase the impact of DBER. Future research and development of change initiatives in DBER should

  - include systematic national surveys or studies of science and engineering teaching practice in each of the disciplines;
  - build on DBER and also on the related fields of faculty development (including the scholarship of teaching and learning), higher education studies, and organizational change;
  - develop and test a range of initiatives aligned with different theories of change;
  - provide empirical data to support claims of success; and

- address a strategic gap in the research by studying new recognition and reward systems designed to encourage research-based improvements in teaching.

The types of studies that the committee recommends would involve different levels and structures of funding than are currently the norm for DBER. Time and money are required to develop and refine measurement instruments and to conduct longitudinal studies; studies that generate sufficient statistical power to make inferences about different student populations; studies of teaching and learning across multiple courses, institutions, or disciplines; and interdisciplinary studies. The committee is confident that with sufficient support, these and the other types of studies on this research agenda have the most potential to build on existing DBER and related research in cognitive science, K-12 science education, psychology, and organizational transformation to generate further insights that can lead to significant improvements in undergraduate science and engineering instruction for all students.

# References

## CHAPTER 1

American Association for the Advancement of Science. (2011). *Vision and change in undergraduate biology education: A call to action.* Washington, DC: American Association for the Advancement of Science.

Association of American Universities. (2011). *Five-year initiative for improving undergraduate STEM education: Discussion draft.* Washington, DC: Association of American Universities.

Bodner, G. (2011). *Status, contributions, and future directions of discipline-based education research: The development of research in chemical education as a field of study.* Paper presented at the Second Committee Meeting on the Status, Contributions, and Future Directions of Discipline-Based Education Research. Available: http://www7. nationalacademies.org/bose/DBER_Bodner_October_Paper.pdf.

Boyer, E.L. (1990). *Scholarship reconsidered: Priorities of the professoriate.* San Francisco, CA: Jossey-Bass.

Brewer, C., and Smith, D. (Eds.). (2011). *Vision and change in undergraduate biology education: A call to action.* Available: http://visionandchange.org/finalreport.

Cooper, M., Grove N., Underwood, S., and Klymkowsky, M. (2010). Lost in Lewis structures: An investigation of student difficulties in developing representational competence. *Journal of Chemical Education, 87*(8), 869-874.

Friedenberg, J., and Silverman, G. (2006). *Cognitive science: An introduction to the study of the mind.* Thousand Oaks, CA: Sage.

Hatch, T. (2005). *Into the classroom: Developing the scholarship of teaching and learning.* San Francisco, CA: Jossey-Bass.

Huber, M., and Hutchings, P. (2005). *The advancement of learning: Building the teaching commons.* San Francisco, CA: Jossey-Bass/Carnegie Foundation for the Advancement of Teaching.

Hutchings, P., Huber, M., and Ciccone, A. (2011). *The scholarship of teaching and learning reconsidered: Institutional integration and impact.* San Francisco, CA: Jossey-Bass.

Mayer, R.E. (2011). *Applying the science of learning to undergraduate science education*. Paper presented at the Third Committee Meeting on Status, Contributions, and Future Directions of Discipline-Based Education Research. Available: http://www7.nationalacademies.org/bose/DBER_Mayer_December_Paper.pdf.

National Academy of Sciences, National Academy of Engineering, and Institute of Medicine. (2011). *Expanding underrepresented minority participation: America's science and technology talent at the crossroads*. Committee on Underrepresented Groups and the Expansion of the Science and Engineering Workforce Pipeline.Committee on Science, Engineering, and Public Policy and Policy and Global Affairs. Washington, DC: The National Academies Press.

National Research Council. (2002). *Scientific research in education*. Committee on Scientific Principles for Education Research, R.J. Shavelson and L. Towne, Eds. Committee on Scientific Principles for Education Research, Center for Education. Division of Behavioral and Social Sciences and Education. Washington, DC: National Academy Press.

National Research Council. (2012). *A framework for K-12 science education practices, cross-cutting concepts, and core ideas*. Committee on a Conceptual Framework for New K-12 Science Education Standard, Board on Science Education. Division of Behavioral and Social Sciences and Education. Washington, DC: The National Academies Press.

Novick, L.R., and Catley, K.M. (in press). Reasoning about evolution's grand patterns: College students' understanding of the tree of life. *American Educational Research Journal*.

President's Council of Advisors on Science and Technology. (2012). *Engage to excel: Producing one million additional college graduates with degrees in science, technology, engineering, and mathematics*. Washington, DC: Executive Office of the President, President's Council of Advisors on Science and Technology.

Project Kaleidoscope. (2011a). *Volume IV: What works, what matters, what lasts: 2004-present: PKAL's online publication*. Washington, DC: Project Kaleidoscope. Available: http://www.pkal.org/collections/VolumeIV.cfm.

Project Kaleidoscope. (2011b). *What works in facilitating interdisciplinary learning in science and mathematics*. Washington, DC: Association of American Colleges and Universities.

Stokes, D.E. (1997). *Pasteur's quadrant: Basic science and technological innovation*. Washington, DC: Brookings Institution Press.

## CHAPTER 2

ABET. (2009). *ABET criteria for evaluating engineering programs*. Baltimore, MD: ABET.

Bailey, J.M. (2011). *Astronomy education research: Developmental history of the field and summary of the literature*. Paper presented at the Second Committee Meeting on the Status, Contributions, and Future Directions of Discipline-Based Education Research. Available: http://www7.nationalacademies.org/bose/DBER_Bailey_Commissioned%20Paper.pdf.

Bauer, C.F., Clevenger, J.V., Cole, R.S., Jones, L.L., Kelter, P.B., Oliver-Hoyo, M.T., and Sawrey, B.A. (2008). Association report, ACS Division of Chemical Education: Hiring and promotion in chemical education. *Journal of Chemical Education*, 85, 898-901.

Beichner, R. (2009). An introduction to physics education research. In C. Henderson and K. Harper (Eds.), *Getting started in PER*. College Park, MD: American Association of Physics Teachers. Available: http://www.per-central.org/items/detail.cfm?ID=8806.

Bloom, B.S. (1956). *Taxonomy of educational objectives, the classification of educational goals* (1st ed.). New York: Longmans, Green.

Bodner, G. (2011). *Status, contributions, and future directions of discipline-based education research: The development of research in chemical education as a field of study.* Paper presented at the Second Committee Meeting on the Status, Contributions, and Future Directions of Discipline-Based Education Research. Available: http://www7.nationalacademies.org/bose/DBER_Bodner_October_Paper.pdf.

Bush, S.D., Pelaez, N.J., Rudd, J.A., Stevens, M.T., Williams, K.S., Allen, D.E., and Tanner, K.D. (2006). On hiring science faculty with education specialties for your science (not education) department. *CBE-Life Sciences Education, 5*(4), 297-305.

Bush, S.D., Pelaez, N.J., Rudd, J.A., Stevens, M.T., Tanner, K.D., and Williams, K.S. (2008). The pipeline: Science faculty with education specialties. *Science, 322*(5909), 1795-1796.

Bush, S.D., Pelaez, N.J., Rudd, J.A., Stevens, M.T., Tanner, K.D., and Williams, K.S. (2011). Investigation of science faculty with education specialties within the largest university system in the United States. *CBE-Life Sciences Education, 10*(1), 25-42.

Clary, R.M., Brzuszek, R.F., and Wandersee, J.H. (2009). Students' geocognition of deep time, conceptualized in an informal educational setting. *Journal of Geoscience Education, 57*(4), 275-285.

Cummings, K. (2011). *A developmental history of physics education research.* Paper presented at the Second Committee Meeting on the Status, Contributions, and Future Directions of Discipline-Based Education Research. Available: http://www7.nationalacademies.org/bose/DBER_Cummings_October_Paper.pdf.

D'Avanzo, C. (2008). Biology concept inventories: Overview, status, and next steps. *Bioscience, 58*(11), 1079-1085.

DeHaan, R.L. (2011). *Education research in the biological sciences: A nine-decade review.* Paper presented at the Second Committee Meeting on the Status, Contributions, and Future Directions of Discipline-Based Education Research. Available: http://www7.nationalacademies.org/bose/DBER_DeHaan_October_Paper.pdf.

Dirks, C. (2011). *The current status and future direction of biology education research.* Paper presented at the Second Committee Meeting on the Status, Contributions, and Future Directions of Discipline-Based Education Research. Available: http://www7.nationalacademies.org/bose/DBER_Dirks_October_Paper.pdf.

Docktor, J.L., and Mestre, J.P. (2011). *A synthesis of discipline-based education research in physics.* Paper presented at the Second Committee Meeting on the Status, Contributions, and Future Directions of Discipline-Based Education Research. Available: http://www7.nationalacademies.org/bose/DBER_Docktor_October_Paper.pdf.

Dodick, J., and Orion, N. (2003). Cognitive factors affecting student understanding of geologic time. *Journal of Research in Science Teaching, 40*(4), 415-442.

Fensham, P.J. (2004). *Defining an identity: The evolution of science education as a field of research.* Boston, MA: Springer.

Feynman, R.P., Leighton, R.B., and Sands, M. (1964). *The Feynman lectures on physics.* Boston, MA: Addison-Wesley.

Finlay, G.C. (1962). The Physical Science Study Committee. *The School Review, 70*(1), 63-81.

French, A.P. (1968). *Special relativity.* New York: Norton.

Hestenes, D., Wells, M., and Swackhamer, G. (1992). Force concept inventory. *The Physics Teacher, 30,* 141-158.

Holton, G. (2003). *The Project Physics Course, then and now.* Cambridge, MA: Harvard University.

Irwin, L. (1970). A brief historical account and introduction to the Earth Science Curriculum Project. *The High School Journal, 53*(4), 241-249.

Kali, Y., and Orion, N. (1996). Spatial abilities of high-school students in the perception of geologic structures. *Journal of Research in Science Teaching, 33,* 369-391.

Karplus, R. (1964). The science curriculum improvement study. *Journal of Research in Science Teaching, 2*, 293-303.

Kastens, K.A., and Ishikawa, T. (2006). Spatial thinking in the geosciences and cognitive sciences. *Geological Society of America Special Papers, 413*, 53-76.

Kern, E.L., and Carpenter, J.R. (1984). Enhancement of student values, interests and attitudes in earth science through a field-oriented approach. *Journal of Geological Education, 32*(5), 299-305.

Kern, E.L., and Carpenter, J.R. (1986). Effect of field activities on student learning. *Journal of Geological Education, 34*(3), 180-183.

Kittel, C., Knight, W.D., and Ruderman, M.A. (1965). *Mechanics: Berkeley physics course: Volume 1.* New York: McGraw-Hill.

Libarkin, J., and Finkelstein, N. (2011). *Seeding DBER: Evaluating the early history of the National Science Foundation's Postdoctoral Fellowships in Science, Mathematics, Engineering and Technology Education (PFSMETE) as a pathway into discipline-based education research.* Paper prepared for the Committee on the Status, Contributions, and Future Directions of Discipline-Based Education Research. Available: http://www7.nationalacademies.org/bose/DBER5_Noah_Finkelstein_Julie_Libarkin.pdf.

Lohmann, J., and Froyd, F. (2011). *Chronological and ontological development of engineering education as a field of scientific inquiry.* Paper presented at the Second Committee Meeting on the Status, Contributions, and Future Directions of Discipline-Based Education Research. Available: http://www7.nationalacademies.org/bose/DBER_Lohmann_Froyd_October_Paper.pdf.

Manduca, C.A, and Mogk, D. (2006). *Earth and mind: How geologists think and learn about the earth.* Boulder, CO: Geological Society of America.

Manduca, C.A., Mogk, D.W., and Stillings, N. (2004). *Bringing research on learning to the geosciences.* Northfield, MN: Carleton College, Science Education Resource Center. Available: http://serc.carleton.edu/files/research_on_learning/ROL0304_2004.pdf.

Manduca, C.A., Mogk, D.W., Tewksbury, B., Macdonald, R.H., Fox, S.P., Iverson, E.R., Kirk, K., McDaris, J., Ormand, C., and Bruckner, M. (2010). SPORE: Science prize for online resources in education. On the cutting edge: Teaching help for geoscience faculty. *Science, 327*(5969), 1095-1096.

Matthews, M.R. (1994). *Science teaching: The role of history and philosophy of science.* New York: Routledge.

Meltzer, D., McDermont, L., Heron, P., Redish, E., and Beichner, R. (2004). *A call to the AAPT executive board and publications committee to expand publication of physics education research articles within the American Journal of Physics.* Available: http://www.ncs u.edu/PER/Articles/CallToAAPT.pdf.

National Academy of Engineering. (2004). *The engineer of 2020: Visions of engineering in the new century.* Washington, DC: The National Academies Press.

National Academy of Sciences, National Academy of Engineering, and Institute of Medicine. (2005). *Facilitating interdisciplinary research.* Committee on Facilitating Interdisciplinary Research. Committee on Science, Engineering, and Public Policy. Washington, DC: The National Academies Press.

National Research Council. (2006). *Learning to think spatially: GIS as a support system in the K-12 curriculum.* Committee on Support for Thinking Spatially: The Incorporation of Geographic Information Science Across the K-12 Curriculum. Geographical Sciences Committee. Board on Earth Sciences and Resources, Division on Earth and Life Studies. Washington, DC: The National Academies Press.

National Science Foundation. (1992). *America's academic future: A report of the presidential young investigator colloquium on U.S. engineering, mathematics, and science education for the year 2010 and beyond.* NSF 91-150. Arlington, VA: National Science Foundation.

National Science Foundation. (1997). *Geoscience education: A recommended strategy.* Report based on August 29-30, 1996, workshop from the Geoscience Education Working Group to the Advisory Committee for Geosciences and the Directorate for Geosciences of the National Science Foundation, Arlington, VA. Available: http://www.nsf.gov/pubs/1997/nsf97171/nsf97171.htm.

Orion, N., and Hofstein, A. (1994). Factors that influence learning during a scientific field trip in a natural environment. *Journal of Research in Science Teaching, 31*(10), 1097-1119.

Orion, N., Hofstein, A., Tamir, P., and Giddings, G.J. (1997). Development and validation of an instrument for assessing the learning environment of outdoor science activities. *Science Education, 81*(2), 161-171.

Petcovic, H.L., Libarkin, J.C., and Baker, K.M. (2009). An empirical methodology for investigating geocognition in the field. *Journal of Geoscience Education, 57*(4), 316-328.

Pfundt, H., and Duit, R. (1988). *Bibliography: Students' alternative frameworks and science education* (2nd ed.). Kiel, Germany: Institute for Science Education.

Rhoten, D., and Parker, A. (2004). Risks and rewards of an interdisciplinary research path. *Science, 306*(5704), 2046.

Rudolph, F. (1990). *The American college and university: A history.* Athens: University of Georgia Press.

Rudolph, J.L. (2002). *Scientists in the classroom: The Cold War reconstruction of American science education.* New York: Palgrave Macmillan.

Semsar, K., Knight, J.K., Birol, G., and Smith, M.K. (2011). The Colorado Learning Attitudes About Science Survey (CLASS) for use in biology. *CBE-Life Sciences Education, 10*(3), 268-278.

Strobel, J., Evangelou, D., Streveler, R.A., and Smith, K.A. (2008). *The many homes of engineering education research: Historical analysis of PhD dissertations.* Paper presented at the Research in Engineering Education Symposium, Davos, Switzerland. Available: http://www.engconfintl.org/8axabstracts/Session%204A/rees08_submission_16.pdf.

Svinicki, M. (2011). *Synthesis of the research on teaching and learning in engineering since the implementation of ABET Engineering Criteria 2000.* Paper presented at the Second Committee Meeting on the Status, Contributions, and Future Directions of Discipline-Based Education Research. Available: http://www7.nationalacademies.org/bose/DBER_Svinicki_October_Paper.pdf.

The Steering Committee of the National Engineering Education Research Colloquies. (2006a). The research agenda for the new discipline of engineering education. *Journal of Engineering Education, 95*(4), 259-261.

The Steering Committee of the National Engineering Education Research Colloquies. (2006b). The National Engineering Education Research Colloquies. *Journal of Engineering Education, 95*(4), 257-258.

## CHAPTER 3

ABET. (2009). *ABET criteria for evaluating engineering programs.* Baltimore, MD: ABET.

Ausubel, D. (2000). *The acquisition and retention of knowledge: A cognitive view.* Boston, MA: Kluwer Academic.

Bailey, J.M. (2011). *Astronomy education research: Developmental history of the field and summary of the literature.* Paper presented at the Second Committee Meeting on the Status, Contributions, and Future Directions of Discipline-Based Education Research. Available: http://www7.nationalacademies.org/bose/DBER_Bailey_Commissioned%20Paper.pdf.

Bailey, J.M., Slater, S.J., and Slater, T.F. (2011). *Conducting astronomy education research: A primer.* New York: W.H. Freeman.

Baily, C., and Finkelstein, N.D. (2011). Interpretive themes in quantum physics: Curriculum development and outcomes. *AIP Conference Proceedings, 1413*, 107-110.

Bassok, M., and Novick, L.R. (2012). Problem solving. In K.J. Holyoak and R.G. Morrison (Eds.), *Oxford handbook of thinking and reasoning* (pp. 413-432). New York: Oxford University Press.

Bhattacharyya, G., and Bodner, G.M. (2005). It gets me to the product: How students propose organic mechanisms. *Journal of Chemical Education, 82*(9), 1402-1407.

Bretz, S.L., and Nakhleh, M.B. (Eds). (2001). Piaget, constructivism, and beyond. *Journal of Chemical Education, 78*(8), 1107.

Brown, J.S., Collins, A., and Duguid, P. (1989). Situated cognition and the culture of learning. *Educational Researcher, 18*(1), 32-42.

Cervato, C., and Frodeman, R. (2012). The significance of geologic time: Cultural, educational, and economic frameworks. *Geological Society of America Special Papers, 486*, 19-27.

Dahl, J., Anderson, S.W., and Libarkin, J.C. (2005). Digging into earth science: Alternative conceptions held by K-12 teachers. *Journal of Science Education, 12*(2), 65-68.

DeLaughter, J.E., Stein, S., and Bain, K.R. (1998). Preconceptions abound among students in an introductory earth science course. *EOS, Transactions of the American Geophysical Union, 79*, 429-432.

Dewey, J. (1916). *Democracy and education.* New York: Macmillan.

Dirks, C. (2011). *The current status and future direction of biology education research.* Paper presented at the Second Committee Meeting on the Status, Contributions, and Future Directions of Discipline-Based Education Research. Available: http://www7.national academies.org/bose/DBER_Dirks_October_Paper.pdf.

Docktor, J.L., and Mestre, J.P. (2011). *A synthesis of discipline-based education research in physics.* Paper presented at the Second Committee Meeting on the Status, Contributions, and Future Directions of Discipline-Based Education Research. Available: http://www7. nationalacademies.org/bose/DBER_Docktor_October_Paper.pdf.

Dodick, J., and Orion, N. (2006). Building an understanding of geological time: A cognitive synthesis of the macro and micro scales of time. *Geological Society of America Special Papers, 413*, 77-93.

Gautier, C., Deutsch, K., and Rebich, S. (2006). Misconceptions about the greenhouse effect. *Journal of Geoscience Education, 54*, 386-395.

Hansen, R., Long, L., and Dellert, T. (2002). *Experiences using flying models in competition and coursework.* Proceedings of the American Society for Engineering Education Zone 1 Conference. West Point, NY: American Society for Engineering Education.

Johnstone, A.H. (1991). Why is science difficult to learn? Things are seldom what they seem. *Journal of Computer Assisted Learning, 7*, 75-83.

Kastens, K.A., Agrawal, S., and Liben, L.S. (2009). How students and field geologists reason in integrating spatial observations from outcrops to visualize a 3-D geological structure. *International Journal of Science Education, 31*(3), 365-393.

Kusnick, J. (2002). Growing pebbles and conceptual prisms: Understanding the source of student misconceptions about rock formation. *Journal of Geosciences Education, 50*(1), 31-39.

Lave, J., and Wenger, E. (1991). *Situated learning. Legitimate peripheral participation.* Cambridge, MA: Cambridge University Press.

Libarkin, J.C., Kurdziel, J.P., and Anderson, S.W. (2007). College student conceptions of geological time and the disconnect between ordering and scale. *Journal of Geoscience Education, 55*(5), 413-422.

Liben, L.S., and Titus, S. (2012). The importance of spatial thinking for geoscience education: Insights from the crossroads of geoscience and cognitive science. *Geological Society of America Special Papers, 86*, 51-70.

Marx, S.M., Weber, E.U., Orlove, B.S., Leiserowitz, A., Krantz, D.K., Roncali, C., and Phillips, J. (2007). Communication and mental processes: Experiential and analytic processing of uncertain climate information. *Global Environmental Change, 17*, 47-58.

Maskall, J., and Stokes, A. (2008). *Designing effective fieldwork for the environmental and natural sciences.* Plymouth, UK: Higher Education Academy Subject Centre for Geography, Earth and Environmental Sciences.

McGourty, J., Scoles, K., and Thorpe, S.W. (2002). *Web-based student evaluation of instruction: Promises and pitfalls.* Paper presented at the Forty-second Annual Forum of the Association for Institutional Research, Toronto, Canada. Available: http://web.augsburg.edu/~krajewsk/educause2004/webeval.pdf.

Mogk, D., and Goodwin, C. (2012). Learning in the field: Synthesis of research on thinking and learning in the geosciences. *Geological Society of America Special Papers, 486*, 131-163.

Mohan, L., Chen, J., and Anderson, C.W. (2009). Developing a multi-year learning progression for carbon cycling in socio-ecological systems. *Journal of Research in Science Teaching, 46*(6), 675-698.

National Academy of Sciences, National Academy of Engineering, and Institute of Medicine. (2005). *Facilitating interdisciplinary research.* Committee on Facilitating Interdisciplinary Research. Committee on Science, Engineering, and Public Policy. Washington, DC: The National Academies Press.

National Research Council. (2003). *Bio2010: Transforming undergraduate education for future research biologists.* Committee on Undergraduate Biology Education to Prepare Research Scientists for the 21st Century. Board on Life Sciences, Division on Earth and Life Studies. Washington, DC: The National Academies Press.

National Research Council. (2006). *Learning to think spatially: GIS as a support system in the K-12 curriculum.* Committee on Support for Thinking Spatially: The Incorporation of Geographic Information Science Across the K-12 Curriculum. Geographical Sciences Committee. Board on Earth Sciences and Resources, Division on Earth and Life Studies. Washington, DC: The National Academies Press.

National Research Council. (2011). *Promising practices in undergraduate science, technology, engineering, and mathematics education: Summary of two workshops.* N. Nielsen, Rapporteur. Planning Committee on Evidence on Selected Innovations in Undergraduate STEM Education. Board on Science Education, Division of Behavioral and Social Sciences and Education. Washington, DC: The National Academies Press.

Orgill, M., and Bodner, G.M. (2006). An analysis of the effectiveness of analogy use in college-level biochemistry textbooks. *Journal of Research in Science Teaching, 43*(10), 1040-1060.

Orgill, M., and Bodner, G. (2007). Locks and keys: An analysis of biochemistry students' use of analogies. *Biochemistry and Molecular Biology Education, 35*(4), 244-254.

Pollock, S., Pepper, R., Chasteen, S., and Perkins, K. (2011). Multiple roles of assessment in upper-division physics course reforms. *AIP Conference Proceedings, 1413*, 307-310.

Rebich, S., and Gautier, C. (2005). Concept mapping to reveal prior knowledge and conceptual change in a mock summit course on global climate change. *Journal of Geoscience Education, 53*(4), 355-365.

Sandi-Urena, S., Cooper, M.M., and Stevens, R.H. (2011). Enhancement of metacognition use and awareness by means of a collaborative intervention. *International Journal of Science Education, 33*(3), 323-340.

Sanger, M.J. (2008). Using inferential statistics to answer quantitative chemical education research questions. In D. Bunce and R. Cole (Eds.), *Nuts and bolts of chemical education research* (pp. 101-133). Washington, DC: American Chemical Society and Oxford University Press.

Smith, T.I., Thompson, J.R., and Mountcastle, D.B. (2010). Addressing student difficulties with statistical mechanics: The Boltzmann factor. *AIP Conference Proceedings, 1289*, 305-308.

Shepardon, D.P., Wee, B., Priddy, M., Schellenberger, L., and Harbor, J. (2007). What is a watershed? Implications for student conceptions for environmental science education and the *National Science Education Standards*. *Science Education, 91*, 555-577.

Slater, S.J., Slater, T.F., and Bailey, J.M. (2011). *Discipline-based education research: A scientist's guide*. New York: W.H. Freeman.

Sterman, J.D., and Sweeney, L.B. (2007). Understanding public complacency about climate change: Adults' mental models of climate change violate conservation of matter. *Climatic Change, 80(3-4)*, 213-238.

Stillings, N. (2012). Complex systems in the geosciences and in geoscience learning. *Geological Society of America Special Papers, 486*, 97-111.

Sundberg, M.D., Dini, M.L., and Li, E. (1994). Decreasing course content improves student comprehension of science and attitudes towards science in freshman biology. *Journal of Research in Science Teaching, 31*, 679-693.

Svinicki, M. (2011). *Synthesis of the research on teaching and learning in engineering since the implementation of ABET engineering criteria 2000*. Paper presented at the Second Committee Meeting on the Status, Contributions, and Future Directions of Discipline-Based Education Research. Available: http://www7.nationalacademies.org/bose/DBER_Svinicki_October_Paper.pdf.

Towns, M.H. (2008). Mixed methods designs in chemical education research. In D. Bunce and R. Cole (Eds.), *Nuts and bolts of chemical education research* (pp. 135-148). Washington, DC: American Chemical Society and Oxford University Press.

Weber, E. (2006). Experience-based and description-based perceptions of long-term risk: Why global warming does not scare us (yet). *Climatic Change, 77*, 103-120.

# CHAPTER 4

Ausubel, D.P., Novak, J.D., and Hanesian, H. (1978). *Educational psychology*. New York: Werbel and Peck.

Bailey, J.M., and Slater, T.F. (2005). A contemporary review of K-16 astronomy education research. In O. Engvold (Ed.), *Highlights of astronomy* (vol. 13, pp. 1029-1031). San Francisco, CA: Astronomical Society of the Pacific.

Bailey, J.M., Johnson, B., Prather, E.E., and Slater, T.F. (2011). Development and validation of the Star Properties Concept Inventory. *International Journal of Science Education*, 1-30.

Bardar, E.M., Prather, B.K., and Slater, T.F. (2006). Development and validation of the Light and Spectroscopy Concept Inventory. *Astronomy Education Review, 5(2)*, 103-113.

Barke, H., Hazari, A., and Yitbarek, S. (2009). *Misconceptions in chemistry: Addressing perceptions in chemical education*. Berlin/Heidelburg: Springer-Verlag.

Besterfield-Sacre, M., Gerchak, J., Lyons, M., Shuman, L.J., and Wolfe, H. (2004). Scoring concept maps: An integrated rubric for assessing engineering education. *Journal of Engineering Education, 93(2)*, 105-115.

Blackwell, W.H., Powell, M.J., and Dukes, G.H. (2003). The problem of student acceptance of evolution. *Journal of Biological Education, 37(2)*, 58-67.

Bodner, G.M. (1991). I have found you an argument: The conceptual knowledge of beginning chemistry graduate students. *Journal of Chemical Education, 68*(5), 385-388.

Brown, D.E., and Clement, J. (1989). Overcoming misconceptions via analogical reasoning: Abstract transfer versus explanatory model construction. *Instructional Science, 18*(4), 237-261.

Cabrera, A.F., Colbeck, C.L., and Terenzini, P.T. (2001). Developing performance indicators for assessing classroom teaching practices and student learning: The case of engineering. *Research in Higher Education, 42*(3), 327-352.

Camp, C.W., Clement, J.J., and Brown, D. (1994). *Preconceptions in mechanics: Lessons dealing with students' conceptual difficulties.* Dubuque, IA: Kendall/Hunt.

Canpolat, N., Pinarbasi, T., and Sözbilir, M. (2006). Prospective teachers' misconceptions of vaporization and vapor pressure. *Journal of Chemical Education, 83*(8), 1237-1242.

Case, J.M., and Fraser, D.M. (1999). An investigation into chemical engineering students' understanding of the mole and the use of concrete activities to promote conceptual change. *International Journal of Science Education, 21*(12), 1237-1249.

Catley, K.M., and Novick, L.R. (2009). Digging deep: Exploring college students' knowledge of macroevolutionary time. *Journal of Research in Science Teaching, 46*(3), 311-332.

Champagne, A.B., Gunstone, R.F., and Klopfer, L.E. (1985). Effecting changes in cognitive structures among physics students. In L. West and L. Pines (Eds.), *Cognitive structure and conceptual change.* Orlando, FL: Academic Press.

Cheek, K.A. (2010) A summary and analysis of twenty-seven years of geoscience conceptions research. *Journal of Geoscience Education, 58*(3), 122-134.

Chi, M.T.H. (2005). Commonsense conceptions of emergent processes: Why some misconceptions are robust. *Journal of the Learning Sciences, 14*(2), 161-199.

Chi, M.T.H. (2008). Three types of conceptual change: Belief revision, mental model transformation, and categorical shift. In S. Vosniadou (Ed.), *International handbook of research on conceptual change* (pp. 61-82). New York: Routledge.

Chinn, C.A., and Brewer, W.F. (1993). The role of anomalous data in knowledge acquisition: A theoretical framework and implications for science instruction. *Review of Educational Research, 63*(1), 1-49.

Clement, J. (1993). Using bridging analogies and anchoring intuitions to deal with students' preconceptions in physics. *Journal of Research in Science Teaching, 30*(10), 1241-1257.

Clement, J. (2008). The role of explanatory models in teaching for conceptual change. In S. Vosniadou (Ed.), *International handbook of research on conceptual change* (pp. 417-452). New York: Routledge.

Clement, J., Brown, D.E., and Zietsman, A. (1989). Not all preconceptions are misconceptions—finding anchoring conceptions for grounding instruction on students' intuitions. *International Journal of Science Education, 11*(5), 554-565.

Cooper, M., Grove N., Underwood, S., and Klymkowsky, M. (2010). Lost in Lewis structures: An investigation of student difficulties in developing representational competence. *Journal of Chemical Education, 87*(8), 869-874.

Cummings, K., Marx, J., Thornton, R., and Kuhl, D. (1999). Evaluating innovation in studio physics. *American Journal of Physics, 67*(Suppl. 1), S38-S44.

Dahl, J., Anderson, S.W., and Libarkin, J.C. (2005). Digging into earth science: Alternative conceptions held by K-12 teachers. *Journal of Science Education, 12*(2), 65-68.

D'Avanzo, C. (2008). Biology concept inventories: Overview, status, and next steps. *Bioscience, 58*(11), 1079-1085.

Dirks, C. (2011). *The current status and future direction of biology education research.* Paper presented at the Second Committee Meeting on the Status, Contributions, and Future Directions of Discipline-Based Education Research. Available: http://www7.nationalacademies.org/bose/DBER_Dirks_October_Paper.pdf.

diSessa, A.A. (1982). Unlearning Aristotelian physics: A study of knowledge-based learning. *Cognitive Science, 6,* 37-75.

diSessa, A.A. (1988). Knowledge in pieces. In G. Forman and P. Putall (Eds.), *Constructivism in the computer age* (pp. 49-70). Hillsdale, NJ: Lawrence Erlbaum Associates.

diSessa, A.A. (2008). A bird's-eye view of the "pieces" vs. "coherence" controversy (from the "pieces" side of the fence). In S. Vosniadou (Ed.), *International handbook of research on conceptual change* (pp. 35-60). New York: Routledge.

diSessa, A.A., Gillespie, N., and Esterly, J. (2004). Coherence vs. fragmentation in the development of the concept of force. *Cognitive Science, 28,* 843-900.

Docktor, J.L., and Mestre, J.P. (2011). *A synthesis of discipline-based education research in physics.* Paper presented at the Second Committee Meeting on the Status, Contributions, and Future Directions of Discipline-Based Education Research. Available: http://www7.nationalacademies.org/bose/DBER_Docktor_October_Paper.pdf.

Englebrecht, A.C., Mintzes, J.J., Brown, L.M., and Kelso, P.R. (2005). Probing understanding in physical geology using concept maps and clinical interviews. *Journal of Geoscience Education, 53*(3), 263-270.

Ericsson, K.A., Krampe, R.Th., and Tesch-Römer, C. (1993). The role of deliberate practice in the acquisition of expert performance. *Psychological Review, 100*(3), 363-406.

Gabel, D.L., Samuel, K.V., and Hunn, D. (1987). Understanding the particulate nature of matter. *Journal of Chemical Education, 64*(8), 695-697.

Henderleiter, J., Smart, R., Anderson, J., and Elian, O. (2001). How do organic chemistry students understand and apply hydrogen bonding? *Journal of Chemical Education, 78*(8), 1126-1130.

Hestenes, D., and Halloun, I. (1995). Interpreting the Force Concept Inventory: A response to March 1995 critique by Huffman and Heller. *Physics Teacher, 33,* 504-506.

Hestenes, D., Wells, M., and Swackhamer, G. (1992). Force Concept Inventory. *The Physics Teacher, 30,* 141-158.

Heywood, J. (2006). Curriculum change and changing the curriculum. In J. Heywood (Ed.), *Engineering education: Research and development in curriculum and instruction* (Ch. 7, pp. 177-197). Hoboken, NJ: John Wiley & Sons.

Hidalgo, A., Fernando, S., and Otero, J. 2004. An analysis of the understanding of geological time by students at secondary and post-secondary level. *International Journal of Science Education, 26*(7), 845-857.

Hornstein, S., Prather, E., English, T., Desch, S., and Keller, J. (2011). Development and testing of the Solar System Concept Inventory. *Bulletin of the American Astronomical Society, 43.*

Hubisz, J. (2001). Report on a study of middle school physical science texts. *The Physics Teacher, 5*(39), 304-309.

Huffman, D., and Heller, P. (1995). What does the Force Concept Inventory actually measure? *Physics Teacher, 33*(2), 138-143.

Jee, B.D., Uttal, D.H., Gentner, D., Manduca, C., Shipley, T.F., Tikoff, B., Ormand, C.J., and Sageman, B. (2010). Commentary: Analogical thinking in geoscience education. *Journal of Geoscience Education, 58*(1), 2-13.

Johnson, J.K., and Reynolds, S.J. (2005). Concept sketches—using student and instructor-generated, annotated sketches for learning, teaching, and assessment in geology courses. *Journal of Geoscience Education, 53*(1), 85-95.

Johnstone, A.H. (1982). Macro and microchemistry. *Chemistry in Britain, 18*(6), 409-410.

Johnstone, A.H. (1991). Why is science difficult to learn? Things are seldom what they seem. *Journal of Computer Assisted Learning, 7,* 75-83.

Kastens, K.A., and Manduca, C.A. (2012). Fostering knowledge integration in geoscience education. *Geological Society of America Special Papers, 486,* 183-206.

King, C.J.H. (2010). An analysis of misconceptions in science textbooks: Earth science in England and Wales. *International Journal of Science Education, 32*(5), 565-601.

Kusnick, J. (2002). Growing pebbles and conceptual prisms: Understanding the source of student misconceptions about rock formation. *Journal of Geoscience Education, 50*(1), 31-39.

Libarkin, J. (2008). *Concept inventories in higher education science.* Paper presented at the National Research Council's Workshop on Linking Evidence to Promising Practices in STEM Undergraduate Education, Washington, DC. Available: http://www7. nationalacademies.org/bose/Libarkin_CommissionedPaper.pdf.

Lindell, R., and Sommer, S. (2003). Using the Lunar Phases Concept Inventory to investigate college students' pre-instructional mental models of lunar phases. *AIP Conference Proceedings, 720*, 73-76.

Lopez, E., Kim, J., Nandagopal, K., Cardin, N., Shavelson, R.J., and Penn, J.H. (2011). Validating the use of concept-mapping as a diagnostic assessment tool in organic chemistry: Implications for teaching. *Chemistry Education Research and Practice, 12*, 133-141.

Marsden, P., and Wright, J. (Eds.). (2010). *Handbook of survey research* (2nd ed.). Bingley, UK: Emerald.

McConnell, D.A., Steer, D.N., Owens, K., Borowski, W., Dick, J., Foos, A., Knott, J.R., Malone, M., McGrew, H., Van Horn, S., Greer, L., and Heaney, P.J. (2006). Using ConcepTests to assess and improve student conceptual understanding in introductory geoscience courses. *Journal of Geoscience Education, 54*(1), 61-68.

McDermott, L.C., and Redish, E.F. (1999). Resource letter: PER-1: Physics education research. *American Journal of Physics, 67*(9), 755-767.

McDermott, L.C., and Shaffer, P.S. (1992). Research as a guide for curriculum development: An example from introductory electricity. Part I: Investigation of student understanding. *American Journal of Physics, 60*, 994-1003.

McNeal, K.S., Miller, H.R., and Herbert, B.E. (2008). Developing nonscience majors' conceptual models of complex earth systems in a physical geology course. *Journal of Geoscience Education, 56*, 201-211.

Minstrell, J. (1989). Teaching science for understanding. In L. Resnick and L. Klopfer (Eds.), *1989 Association for Supervision and Curriculum Development yearbook: Toward the thinking curriculum: Current cognitive research.* Alexandria, VA: Association for Supervision and Curriculum Development.

Mohan, L., Chen, J., and Anderson, C.W. (2009). Developing a multi-year learning progression for carbon cycling in socio-ecological systems. *Journal of Research in Science Teaching, 46*(6), 675-698.

Montfort, D., Brown, S., and Pollock, D. (2009). An investigation of students' conceptual understanding in related sophomore to graduate-level engineering and mechanics courses. *Journal of Engineering Education, 98*(2), 111-129.

Morabito, N.P., Catley, K.M., and Novick, L.R. (2010). Reasoning about evolutionary history: Postsecondary students' knowledge of most recent common ancestry and homoplasy. *Journal of Biological Education, 44*(4), 166-174.

Mulford, D.R., and Robinson, W.R. (2002). An inventory for alternate conceptions among first-semester general chemistry students. *Journal of Chemical Education, 79*(6), 739-744.

Muryanto, S. (2006). Concept mapping: An interesting and useful tool for chemical engineering laboratories. *International Journal of Engineering Education, 22*(5), 979-985.

Nakhleh, M.B. (1992). Why some students don't learn chemistry: Chemical misconceptions. *Journal of Chemical Education, 69*(3), 191-196.

National Academy of Sciences, National Academy of Engineering, and Institute of Medicine. (2005). *Facilitating interdisciplinary research.* Committee on Facilitating Interdisciplinary Research. Committee on Science, Engineering, and Public Policy. Washington, DC: The National Academies Press.

National Research Council. (1999). *How people learn: Bridging research and practice*. M.S. Donovan, J.D. Bransford, and J.W. Pellegrino (Eds.). Committee on Learning Research and Educational Practice, Commission on Behavioral and Social Sciences and Education. Washington, DC: National Academy Press.

National Research Council. (2007). *Taking science to school: Learning and teaching science in grades K-8*. Committee on Science Learning, Kindergarten Through Eighth Grade, R.A. Duschl, H.A. Schweingruber, and A.W. Shouse (Eds.). Board on Science Education, Center for Education. Division of Behavioral and Social Sciences and Education Washington, DC: The National Academies Press.

Nicoll, G. (2003). A qualitative investigation of undergraduate chemistry students' macroscopic interpretations of the submicroscopic structures of molecules. *Journal of Chemical Education, 80*, 205-213.

Novak, J.D., and Gowin, D.B. (1984). *Learning how to learn*. New York: Cambridge University Press.

O'Brien, R.J., Lau, L.F., and Huganir, R.L. (1998) Molecular mechanisms of glutamate receptor clustering at excitatory synapses. *Current Opinion Neurobiology, 8*, 364-369.

Oreskes, N., Conway, E.M., and Shindell, M. (2008). From Chicken Little to Dr. Pangloss: William Nierenberg, global warming, and the social deconstruction of scientific knowledge. *Historical Studies in the Natural Sciences, 38*(1), 109-152.

Orgill, M., and Sutherland, A. (2008). Undergraduate chemistry students' perceptions of and misconceptions about buffers and buffer problems. *Chemistry Education Research and Practice, 9*(2), 131-143.

Pelaez, N.J., Boyd, D.D., Rojas, J.B., and Hoover, M.A. (2005). Prevalence of blood circulation misconceptions among prospective elementary teachers. *Advances in Physiology Education, 29*(3), 172-181.

Pintrich, P.R. (1999). Motivational beliefs as resources for and constraints on conceptual change. In W. Schnotz, S. Vosniadou, and M. Carretero (Eds.), *New perspectives on conceptual change*. Oxford, UK: Pergamon Press.

Plummer, J.D., and Krajcik, J. (2010). Building a learning progression for celestial motion: Elementary levels from an earth-based perspective. *Journal of Research in Science Teaching, 47*(7), 768-787.

Posner, G.J., Strike, K.A., Hewson, P.W., and Gertzog, W.A. (1982). Accommodation of a scientific conception: Toward a theory of conceptual change. *Science Education, 66*(2), 211-227.

Prince, M.J., Vigeant, M.A.S., and Nottis, K. (2009). A preliminary study on the effectiveness of inquiry-based activities for addressing misconceptions of undergraduate engineering students. *Education for Chemical Engineers, 4*(2), 29-41.

Rebich, S., and Gautier, C. (2005). Concept mapping to reveal prior knowledge and conceptual change in a mock summit course on global climate change. *Journal of Geoscience Education, 53*(4), 355-365.

Reed-Rhoads, T., and Imbrie, P.K. (2008). *Concept inventories in engineering education*. Paper presented at the National Research Council's Workshop on Linking Evidence to Promising Practices in STEM Undergraduate Education, Washington, DC. Available: http://www7.nationalacademies.org/bose/Reed_Rhoads_CommissionedPaper.pdf.

Rushton, G.T., Hardy, R.C., Gwaltney, K.P., and Lewis, S.E. (2008). Alternative conceptions of organic chemistry topics among fourth year chemistry students. *Chemistry Education Research and Practice, 9*(2), 122-130.

Schneps, M.H., and Sadler, P.M. (1987). *A private universe*. Available: http://www.learner.org/resources/series28.html.

Schwarz, C.V., Reiser, B.J., Davis, E.A., Kenyon, L., Achér, A., Fortus, D., Shwartz, Y., Hug, B., and Krajcik, J. (2009). Developing a learning progression for scientific modeling: Making scientific modeling accessible and meaningful for learners. *Journal of Research in Science Teaching, 46*(6), 632-368.

Segalas, J., Ferrer-Balas, D., and Mulder, K. (2008). Conceptual maps: Measuring learning processes of engineering students concerning sustainable development. *European Journal of Engineering Education, 33*(3), 297-306.

Sinatra, G.M., Southerland, S.A., McConaughy, F., and Demastes, J.W. (2003). Intentions and beliefs in students' understanding and acceptance of biological evolution. *Journal of Research in Science Teaching, 40*(5), 510-528.

Slater, S.J., Slater, T.F., and Bailey, J.M. (2011). *Discipline-based education research: A scientist's guide.* New York: W.H. Freeman.

Smith, M.K., Wood, W.B., and Knight, J.K. (2008). The genetics concept assessment: A new concept inventory for gauging student understanding of genetics. *CBE-Life Sciences Education, 7*(4), 422-430.

Sokoloff, D.R., and Thornton, R.K. (1997). Using interactive lecture demonstrations to create an active learning environment. *The Physics Teacher, 35*(6), 340-347.

Sözbilir, M. (2002). Turkish chemistry undergraduate students' misunderstandings of Gibbs free energy. *University Chemistry Education, 6*(2), 73-83.

Sözbilir, M. (2004). What makes physical chemistry difficult? Perceptions of Turkish chemistry undergraduates and lecturers. *Journal of Chemical Education, 81*(4), 573-578.

Sözbilir, M., and Bennett, J.M. (2006). Turkish prospective chemistry teachers' misunderstandings of enthalpy and spontaneity. *Chemical Educator, 11*(5), 355-363.

Stevens, R., O'Connor, K., Garrison, L., Jocuns, A., and Amos, D.M. (2008). Becoming an engineer: Toward a three-dimensional view of engineering learning. *Journal of Engineering Education, 97*(3), 355-368.

Svinicki, M. (2011). *Synthesis of the research on teaching and learning in engineering since the implementation of ABET engineering criteria 2000.* Paper presented at the Second Committee Meeting on the Status, Contributions, and Future Directions of Discipline-Based Education Research. Available: http://www7.nationalacademies.org/bose/DBER_Svinicki_October_Paper.pdf.

Taber, K.S. (2001) The mismatch between assumed prior knowledge and the learner's conceptions: a typology of learning impediments, *Educational Studies, 27*(2), 159-171.

Talanquer, V. (2002). Minimizing misconceptions: Tools for identifying patterns of reasoning. *The Science Teacher, 69*, 46-49.

Talanquer, V. (2006). Common-sense chemistry: A model for understanding students' alternative conceptions. *Journal of Chemical Education, 83*, 811.

Tanner, K., and Allen, D. (2005). Approaches to biology teaching and learning: Understanding the wrong answers—teaching toward conceptual change. *Cell Biology Education, 4*(2), 112-117.

Taraban, R., Anderson, E.E., DeFinis, A., Brown, A.G., Weigold, A., and Sharma, M.P. (2007). First steps in understanding engineering students' growth of conceptual and procedural knowledge in an interactive learning context. *Journal of Engineering Education, 96*(1), 57-68.

Teed, R., and Slattery, W. (2011). Changes in geologic time understanding in a class for preservice teachers, *Journal of Geoscience Education, 59*, 151-162.

Treagust, D. (1988). Teaching science for conceptual change: Theory and practice. In S. Vosniadou (Ed.), *International handbook of research on conceptual change* (pp. 629-646). New York: Routledge.

Trend, R. (2000). Conceptions of geological time among primary teacher trainees, with reference to their engagement with geoscience, history, and science. *International Journal of Science Education, 22*(5), 539-555.

Turns, J., Atman, C., Adams, R., and Barker, T. (2005). Research on engineering student knowing: Trends and opportunities. *Journal of Engineering Education, 94*(1), 27-41.

Vosniadou, S. (2008a). Conceptual change research: An introduction. In S. Vosniadou (Ed.), *International handbook of research on conceptual change* (pp. xii-xxviii). New York: Routledge.

Vosniadou, S. (Ed.). (2008b). *International handbook of research on conceptual change.* New York: Routledge.

Vosniadou, S., Vamvakoussi, X., and Skopeliti, I. (2008). The framework theory approach to the problem of conceptual change. In S. Vosniadou (Ed.), *International handbook of research on conceptual change* (pp. 3-34). New York: Routledge.

Wandersee, J.H., Mintzes, J.J., and Novak, J.D. (1994). Research on alternative conceptions in science. In D. Gabel (Ed.), *Handbook of research in science teaching and learning* (pp. 177-210). New York: Simon & Schuster Macmillan.

Yezierski, E.J., and Birk, J.P. (2006). Misconceptions about the particulate nature of matter. Using animations to close the gender gap. *Journal of Chemical Education, 83*(6), 954-960.

Zeilik, M. (2002). Birth of the Astronomy Diagnostic Test: Prototest evolution. *Astronomy Education Review, 1*(2), 46.

Zeilik, M., and Bisard, W. (2000). Conceptual change in introductory-level astronomy courses. *Journal of College Science Teaching, 29*(4), 229-232.

Zeilik, M., Schau, C., Mattern, N., Hall, S., Teague, K.W., and Bisard, W. (1997). Conceptual astronomy: A novel model for teaching postsecondary science courses. *American Journal of Physics, 65*(10), 987-996.

Zeilik, M., Mattern, N., and Schau, C. (1999). Conceptual astronomy II. Replicating conceptual gains, probing attitude changes across three semesters. *American Journal of Physics, 67*(10), 923-927.

Zimrot, R., and Ashkenazi, G. (2007). Interactive lecture demonstrations: A tool for exploring and enhancing conceptual change. *Chemistry Education Research and Practice, 8,* 197-211.

## CHAPTER 5

ABET. (1995). *Vision for change: A summary report of the ABET/NSF/industry workshops.* Baltimore, MD: ABET.

Abraham, M., Varghese, V., and Tang, H. (2010). Using molecular representations to aid student understanding of stereochemical concepts. *Journal of Chemical Education, 87*(12), 1425-1429.

Ainsworth, S., Prain, V., and Tytler, R. (2011). Drawing to learn science. *Science, 333,* 1096-1097.

Aldahmash, A.H., and Abraham, M.R. (2009). Kinetic versus static visuals for facilitating college students' understanding of organic reaction mechanisms in chemistry. *Journal of Chemical Education, 86*(12), 1442-1446.

Allard, F., and Starkes, J. (1991). Motor-skill experts in experts in sports, dance and other domains. In K.A. Ericsson and H.G. Smith (Eds.), *Toward a general theory of expertise* (pp. 126-152). Cambridge, UK: Cambridge University Press.

Anderson, W.L., Mitchell, S.M., and Osgood, M.P. (2008). Gauging the gaps in student problem-solving skills: Assessment of individual and group use of problem-solving strategies using online discussions. *CBE-Life Sciences Education, 7*, 254-262.

Anderson, W.L., Sensibaugh, C.A., Osgood, M.P., and Mitchell, S.M. (2011). What really matters: Assessing individual problem-solving performance in the context of biological sciences. *International Journal for the Scholarship of Teaching and Learning, 5*(1), 1-20.

Anzai, Y. (1991). *Learning and use of representations for physics expertise.* In K. Ericson and J. Smith (Eds.), *Toward a general theory of expertise: Prospects and limits* (pp. 64-90). Cambridge, UK: Cambridge University Press.

Atman, C., Kilgore, D., and McKenna, A. (2008). Characterizing design learning: A mixed methods study of engineering designers' use of language. *Journal of Engineering Education, 97*(3), 309-326.

Ault, C.R., Jr. (1994). Research on problem-solving: Earth science. In D. Gabel (Ed.), *Handbook of research on science teaching and learning.* New York: MacMillan.

Baddeley, A. (2007). *Working memory, thought, and action.* Oxford, UK: Oxford University Press.

Bagno, E., and Eylon, B. (1997). From problem solving to a knowledge structure: An example from the domain of electromagnetism. *The American Journal of Physics, 65*(8), 726-736.

Baillie, C., Goodhew, P., and Skryabina, E. (2006). Threshold concepts in engineering education: Exploring potential blocks in student understanding. *International Journal of Engineering Education, 22*(5), 955-962.

Barrows, H.S. (1986). A taxonomy of problem-based learning methods. *Medical Education, 20*, 481-486.

Bassok, M., and Novick, L.R. (2012). Problem solving. In K.J. Holyoak and R.G. Morrison (Eds.), *Oxford handbook of thinking and reasoning* (pp. 413-432). New York: Oxford University Press.

Bassok, M., and Olseth, K.L. (1995). Object-based representations: Transfer between cases of continuous and discrete models of change. *Journal of Experimental Psychology: Learning, Memory, and Cognition, 21*(6), 1522-1538.

Bassok, M., Chase, V.M., and Martin, S.A. (1998). Adding apples and oranges: Alignment of semantic and formal knowledge. *Cognitive Psychology, 35*(2), 99-134.

Beichner, R.J. (1994). Testing student interpretation of kinematics graphs. *American Journal of Physics, 62*(8), 750-762.

Bhattacharyya, G., and Bodner, G.M. (2005). It gets me to the product: How students propose organic mechanisms. *Journal of Chemical Education, 82*(9), 1402-1407.

Bodner, G.M. (2004). Twenty years of learning: How to do research in chemical education. *Journal of Chemical Education, 81*(5), 618-628.

Bodner, G.M., and Herron, J.D. (2002). Problem solving in chemistry. In J.K. Gilbert (Ed.), *Chemical education: Research-based practice.* Dordecht, The Netherlands: Kluwer Academic.

Bodner, G.M., and McMillan, T.L.B. (1986). Cognitive restructuring as an early stage in problem-solving. *Journal of Research in Science Teaching, 23*(8), 727-737.

Bransford, J., Vye, N., and Bateman, H. (2002). Creating high-quality learning environments: Guidelines from research on how people learn. In National Research Council, *The knowledge economy and postsecondary education: Report of a workshop* (pp. 159-197). Committee on the Impact of the Changing Economy of the Education System. P.A. Graham and N.G. Stacey (Eds.). Center for Education, Division of Behavioral and Social Sciences and Education. Washington, DC: National Academy Press.

Brookes, D.T., and Etkina, E. (2007). Using conceptual metaphor and functional grammar to explore how language used in physics affects student learning. *Physical Review Special Topics—Physics Education Research, 3*(1), 010105-1-010105-16.

Brown, J.S., Collins, A., and Duguid, P. (1989). Situated cognition and the culture of learning. *Educational Researcher, 18*(1), 32-42.

Bunce, D.M., and Gabel, D. (2002). Differential effects on the achievement of males and females of teaching the particulate nature of chemistry. *Journal of Research in Science Teaching, 39*(10), 911-927.

Bunce, D.M., and Heikkinen, H. (1986). The effects of an explicit problem-solving approach on mathematical chemistry achievement. *Journal of Research in Science Teaching, 23*(1), 11-20.

Camacho, M., and Good, R. (1989). Problem solving and chemical equilibrium: Successful versus unsuccessful performance. *Journal of Research in Science Teaching, 26*(3), 251-272.

Carlson, C.A. (1999). Field research as a pedagogical tool for learning hydrogeochemistry and scientific-writing skills. *Journal of Geoscience Education, 47*(2), 150-157.

Catley, K.M., and Novick, L.R. (2008). Seeing the wood for the trees: An analysis of evolutionary diagrams in biology textbooks. *Bioscience, 58*(10), 976-987.

Chandrasegaran, A.L., Treagust, D.F., Waldrip, B.G., and Chandrasegaran, A. (2009). Students' dilemmas in reaction stoichiometry problem solving: Deducing the limiting reagent in chemical reactions. *Chemistry Education Research and Practice, 10*(1), 14-23.

Chase, W.G., and Simon, H.A. (1973). The mind's eye in chess. In W.G. Chase (Ed.), *Visual information processing* (pp. 215-281). New York: Academic Press.

Cheville, A. (2010). Transformative experiences: Scaffolding design learning through the Vygotsky cycle. *International Journal of Engineering Education, 26*(4), 760-770.

Chi, M.T.H. (2006). Two approaches to the study of experts' characteristics. In K.A. Ericsson, N. Charness, R.R. Hoffman, and P.J. Feltovich (Eds.), *The Cambridge handbook of expertise and expert performance* (pp. 21-30). New York: Cambridge University Press.

Chi, M.T.H., Feltovich, P.J., and Glaser, R. (1981). Categorization and representation of physics problems by experts and novices. *Cognitive Science, 5*(2), 121-152.

Chi, M.T.H., Glaser, R., and Rees, E. (1982). Expertise in problem solving. In R. Sternberg (Ed.), *Advances in the psychology of human intelligence* (vol. 1, pp. 7-76). Hillsdale, NJ: Lawrence Erlbaum Associates.

Chi, M.T.H., Bassok, M., Lewis, M.W., Reimann, P., and Glaser, R. (1989). Self-explanations: How students study and use examples in learning to solve problems. *Cognitive Science, 13*(2), 145-182.

Clement, J. (2009). Creative model construction in scientists and students. *The role of imagery, analogy and mental simulations.* Dordrecht, The Netherlands: Springer.

Cohen, E., Mason, A., Singh, C., and Yerushalmi, E. (2008). Identifying differences in diagnostic skills between physics students: Students' self-diagnostic performance given alternative scaffolding. *AIP Conference Proceedings, 1064,* 99-102.

Collins, A., Brown, J.S., and Newman, S.E. (1989). Cognitive apprenticeship: Teaching the crafts of reading, writing, and mathematics. In L.B. Resnick (Ed.), *Knowing, learning, and instruction: Essays in honor of Robert Glaser* (pp. 453-494). Hillsdale, NJ: Lawrence Erlbaum Associates.

Connor, C. (2009). Field glaciology and earth systems science: The Juneau Icefield Research Program (JIRP), 1946-2008. *Geological Society of America Special Papers, 461,* 173-184.

Cooper, M.M., Cox, C.T., Jr., Nammouz, M., Case, E., and Stevens, R.J. (2008). An assessment of the effect of collaborative groups on students' problem-solving strategies and abilities. *Journal of Chemical Education, 85*(6), 866-872.

Cooper, M., Grove, N., Underwood, S., and Klymkowsky, M. (2010). Lost in Lewis structures: An investigation of student difficulties in developing representational competence. *Journal of Chemical Education, 87*(8), 869-874.

Cordray, D.S., Harris, T.R., and Klein, S. (2009). A research synthesis of the effectiveness, replicability, and generality of the VaNTH: Challenge-based instructional modules in bioengineering. *Journal of Engineering Education, 98*(4), 335-348.

Coughlin, L.D., and Patel, V.L. (1987). Processing of critical information by physicians and medical students. *Journal of Medical Education, 62*(10), 818-828.

Cummings, K., Marx, J., Thornton, R., and Kuhl, D. (1999). Evaluating innovation in studio physics. *American Journal of Physics, 67*(7, Suppl. 1), S38-S44.

de Jong, T., and Ferguson-Hessler, M.G.M. (1986). Cognitive structures of good and poor novice problem solvers in physics. *Journal of Educational Psychology, 78,* 279-288.

de Wet, A., Manduca, C., Wobus, R.A., and Bettison-Varga, L. (2009). Twenty-two years of undergraduate research in the geosciences—the Keck experience. *Geological Society of America Special Papers, 461,* 163-172.

Dirks, C. (2011). *The current status and future direction of biology education research.* Paper presented at the Second Committee Meeting on the Status, Contributions, and Future Directions of Discipline-Based Education Research. Available: http://www7.national academies.org/bose/DBER_Dirks_October_Paper.pdf.

Docktor, J.L., and Heller, K. (2009). Assessment of student problem solving processes. *AIP Conference Proceedings, 1179,* 133-136.

Docktor, J.L., and Mestre, J.P. (2011). *A synthesis of discipline-based education research in physics.* Paper presented at the Second Committee Meeting on the Status, Contributions, and Future Directions of Discipline-Based Education Research. Available: http://www7.nationalacademies.org/bose/DBER_Docktor_October_Paper.pdf.

Duch, B.J. (1997). Problem-based learning in physics: Making connections with the real world. *AIP Conference Proceedings, 399,* 557-559.

Dufour-Janvier, B., Bednarz, N., and Belanger, M. (1987). Pedagogical considerations concerning the problem of representation. In C. Janvier (Ed.), *Problems of representation in the teaching and learning of mathematics* (pp. 109-122). Hillsdale, NJ: Lawrence Erlbaum Associates.

Dufresne, R.J., Gerace, W.J., Hardiman, P.T., and Mestre, J.P. (1992). Constraining novices to perform expert-like problem analyses: Effects on schema acquisition. *Journal of the Learning Sciences, 2,* 307-331.

Dutrow, B.L. (2007). Visual communication: Do you see what I see? *Elements, 3*(2), 119-126.

Dutson, A., Todd, R., Magelby, S., and Sorenson, C. (1997). A review of literature on teaching engineering design through project-oriented capstone courses. *Journal of Engineering Education, 86*(1), 17-28.

Dym, C.L., Agogino, A.M., Eris, O., Frey, D.D., and Leifer, L.J. (2005). Engineering design thinking, teaching, and learning. *Journal of Engineering Education, 94*(1), 103-119.

Eastman, C. (2001). New directions in design cognition: Studies of representation and recall. In C. Eastman, W. Newstetter, and M. McCracken (Eds.), *Design knowing and learning: Cognition in design education* (pp. 147-198). Oxford, UK: Elsevier Science.

Egan, D.E., and Schwartz, B.J. (1979). Chunking in recall of symbolic drawings. *Memory and Cognition, 7*(2), 149-158.

Ericsson, K.A., and Simon, H.A. (1980). Verbal reports as data. *Psychological Review, 87,* 215-251.

Ericsson, K.A., and Simon, H.A. (1993). *Protocol analysis: Verbal reports as data* (rev. ed.). Cambridge, MA: MIT Press.

Eylon, B., and Reif, F. (1984). Effects of knowledge organization on task performance. *Cognition and Instruction, 1,* 5-44.

Fay, M.E., Grove, N.P., Towns, M.H., and Bretz, S.L. (2007). A rubric to characterize inquiry in the undergraduate chemistry laboratory. *Chemistry Education Research and Practice, 8*(2), 212-219.

Ferrini-Mundy, J., and Graham, K. (1994). Research in calculus learning: Understanding of limits, derivatives, and integrals. In J. Kaput and E. Dubinsky (Eds.), *Research issues in undergraduate mathematics learning: Preliminary analyses and results* (pp. 31-35). Washington, DC: Mathematical Association of America.

Finegold, M., and Mass, R. (1985). Differences in the process of solving physics problems between good problem solvers and poor problem solvers. *Research in Science and Technology Education, 3,* 59-67.

Fuhrman, O., and Boroditsky, L. (2010). Cross-cultural differences in mental representations of time: Evidence from an implicit non-linguistic task. *Cognitive Science, 34,* 1430-1451.

Fuller, I., Edmondson, S., France, D., Higgitt, D., and Ratinen, I. (2006). International perspectives on the effectiveness of geography fieldwork for learning. *Journal of Geography in Higher Education, 30*(1), 89-101.

Gabel, D., and Bunce, D. (1994). Research on problem solving: Chemistry. In D. Gabel (Ed.), *Handbook of research on science teaching and learning* (pp. 301-326). New York: Macmillan.

Gerson, H.B.P., Sorby, S.A., Wysocki, A., and Baartmans, B.J. (2001). The development and assessment of multimedia software for improving 3-D visualization skills. *Computer Applications in Engineering Education, 9,* 105-113.

Gertner, A.S., and VanLehn, K. (2000). Andes: A coached problem-solving environment for physics. In G. Gautheier, C. Frasson, and K. VanLehn (Eds.), *Intelligent tutoring systems: 5th international conference, ITS 2000* (pp. 133-142). New York: Springer.

Gilbert, J.K., and Treagust, D.F. (Eds.). (2009). *Multiple representations in chemical education.* New York: Springer.

Glasgow, J., Narayanan, N.H., and Chandrasekaran, B. (1995). *Diagrammatic reasoning: Cognitive and computational perspectives* (pp. 303-338). Menlo Park, CA: AAAI Press.

Gobet, F., and Simon, H.A. (1996). The roles of recognition processes and look-ahead search in time-constrained expert problem solving: Evidence from grand-master-level chess. *Psychological Science, 7*(1), 52-55.

Goldberg, F.M., and Anderson, J.H. (1989). Student difficulties with graphical representations of negative values of velocity. *Physics Teacher, 27*(4), 254-260.

Gonzales, D., and Semken, S. (2009). Using field-based research studies to engage undergraduate students in igneous and metamorphic petrology: A comparative study of pedagogical approaches and outcomes. *Geological Society of America Special Papers, 461,* 205-221.

Goodwin, C. (1994). Professional vision. *American Anthropologist, 96*(3), 606-633.

Hardiman, P.T., Dufresne, R., and Mestre, J.P. (1989). The relation between problem categorization and problem solving among experts and novices. *Memory and Cognition, 17*(5), 627-638.

Harris, M.A., Peck, R.F., Colton, S., Morris, J., Chaibub Neto, E., and Kallio, J. (2009). A combination of hand-held models and computer imaging programs helps students answer oral questions about molecular structure and function: A controlled investigation of student learning. *CBE-Life Sciences Education, 8,* 29-43.

Hayes, J.R., and Simon, H.A. (1977). Psychological differences among problem isomorphs. In N.J. Castellan, D.B. Pisoni, and G.R. Potts (Eds.), *Cognitive theory* (vol. 2, pp. 21-44). Hillsdale, NJ: Lawrence Erlbaum Associates.

Hegarty, M. (1992). Mental animation: Inferring motion from static diagrams of mechanical systems. *Journal of Experimental Psychology: Learning, Memory and Cognition, 18*(5), 1084-1102.

Hegarty, M. (2011). *The role of spatial thinking in undergraduate science education.* Paper presented at the Third Committee Meeting on Status, Contributions, and Future Directions of Discipline-Based Education Research. Available: http://www7.nationalacademies.org/bose/DBER_Hegarty_December_Paper.pdf.

Hegarty, M., and Sims, V.K. (1994). Individual differences in mental animation during mechanical reasoning. *Memory and Cognition, 22*(4), 411-430.

Hegarty, M., Carpenter, P.A., and Just, M.A. (1991). Diagrams in the comprehension of scientific texts. In R. Barr, M.L. Kamil, P. Mosenthal, and P.D. Pearson (Eds.), *Handbook of reading research.* New York: Longman.

Hegarty, M., Kriz, S., and Cate, C. (2003). The roles of mental animations and external animations in understanding mechanical systems. *Cognition and Instruction, 21*(4), 209-249.

Hegarty, M., Crookes, R., Dara-Abrams, D., and Shipley, T.F. (2010). Do all science disciplines rely on spatial abilities? Preliminary evidence from self-report questionnaires. In C. Hölscher, T.E. Shipley, M.O. Belardanelli, J.A. Bateman, and N.S. Newcombe (Eds.), *Proceedings of spatial cognition 2010* (pp. 85-94). Berlin, Germany: Springer.

Heller, J., and Reif, F. (1984). Prescribing effective human problem-solving processes: Problem description in physics. *Cognition and Instruction, 1*(2), 177-216.

Heller, K., and Heller, P. (2000). *Competent problem solver—calculus version.* New York: McGraw-Hill.

Heller, P., and Hollabaugh, M. (1992). Teaching problem-solving through cooperative grouping. Part 2: Designing problems and structuring groups. *American Journal of Physics, 60*(7), 637-644.

Heller, P., Keith, R., and Anderson, S. (1992). Teaching problem solving through cooperative grouping. Part 1: Group versus individual problem solving. *American Journal of Physics, 60*(7), 627-636.

Henderleiter, J., Smart, R., Anderson, J., and Elian, O. (2001). How do organic chemistry students understand and apply hydrogen bonding? *Journal of Chemical Education, 78*(8), 1126-1130.

Herron, J.D., and Greenbowe, T.J. (1986). What can we do about Sue: A case study of competence. *Journal of Chemical Education, 63*, 526-531.

Hoellwarth, C., Moelter, M.J., and Knight, R.D. (2005). A direct comparison of conceptual learning and problem-solving ability in traditional and studio style classrooms. *American Journal of Physics, 73*(5), 459-462.

Hsu, L., and Heller, K. (2009). *Computer problem-solving coaches.* Proceedings of the National Association for Research in Science Teaching 82nd International Conference, Garden Grove, CA.

Hsu, L., Brewe, E., Foster, T.M., and Harper, K.A. (2004). Resource letter RPS-1: Research in problem solving. *American Journal of Physics, 72*(9), 1147-1156.

Hundhausen, C., Agarwal, P., Zollars, R., and Carter, A. (2011). The design and experimental evaluation of a scaffolded software environment to improve engineering students' disciplinary problem-solving skills. *Journal of Engineering Education, 100*(3), 574-603.

Hurley, S.M., and Novick, L.R. (2010). Solving problems using matrix, network, and hierarchy diagrams: The consequences of violating construction conventions. *Quarterly Journal of Experimental Psychology, 63*(2), 275-290.

Hynd, C., McWhorter, J.Y., Phares, V.L., and Suttles, C.W. (1994). The role of instructional variables in conceptual change in high school physics topics. *Journal of Research in Science Teaching, 31*(9), 933-946.

Isaak, M.I., and Just, M.A. (1995). Constraints on the processing of rolling motion: The curtate cycloid illusion. *Journal of Experimental Psychology: Human Perception and Performance, 21*(6), 1391-1408.

Ishikawa, T., Barnston, A.G., Kastens, K.A., Louchouarn, P., and Ropelewski, C.F. (2005). Climate forecast maps as a communication and decision-support tool: An empirical test with prospective policy makers. *Cartography and Geographic Information Science, 32*(1), 3-16.

Ishikawa, T., Barnston, A.G., Kastens, K.A., and Louchouarn, P. (2011). Understanding, evaluation, and use of climate forecast data by environmental policy students. *Geological Society of America Special Papers, 474*, 153-170.

Johnson, D.W., Johnson, R.T., and Smith, K.A. (1991). *Cooperative learning: Increasing college faculty instructional productivity.* ASHE-ERIC Higher Education Report No. 4. Washington, DC: The George Washington University School of Education and Human Development.

Johnstone, A.H. (1991). Why is science difficult to learn? Things are seldom what they seem. *Journal of Computer Assisted Learning, 7*, 75-83.

Johnstone, A.H., and El-Banna, H. (1986). Capacities, demands and processes: A predictive model for science education. *Education in Chemistry, 23*, 80-84.

Kali, Y., and Orion, N. (1996). Spatial abilities of high-school students in the perception of geologic structures. *Journal of Research in Science Teaching, 33*(4), 369-391.

Kastens, K. (2009, October). *Synthesis of research on thinking and learning in the geosciences: Developing representational competence.* Paper presented at the annual meeting of the Geological Society of America. Abstract available: https://gsa.confex.com/gsa/2009AM/finalprogram/abstract_165183.htm.

Kastens, K. (2010). Object and spatial visualization in geosciences. *Journal of Geoscience Education, 58*(2), 52-57.

Kastens, K.A., and Manduca, C.A. (2012). Fostering knowledge integration in geoscience education. *Geological Society of America Special Papers, 486*, 183-206.

Kellman, P.J. (2000). An update on Gestalt psychology. In B. Landau, J. Sabini, J. Jonides, and E. Newport (Eds.), *Perception, cognition, and language: Essays in honor of Henry and Lila Gleitman.* Cambridge, MA: MIT Press.

Kelly, R.M., and Jones, L.L. (2008). Investigating students' ability to transfer ideas learned from molecular animations of the dissolution process. *Journal of Chemical Education, 85*(2), 303-309.

Kindfield, A.C.H. (1993/1994). Biology diagrams: Tools to think with. *Journal of the Learning Sciences, 3*, 1-36.

Klein, G.A., Orasanu, J., Calderwood, R., and Zsambok, C.E. (Eds.). (1993). *Decision making in action: Models and methods.* Norwood, NJ: Ablex.

Kohl, P.B., and Finkelstein, N.D. (2005). Student representational competence and self-assessment when solving physics problems. *Physical Review Special Topics—Physics Education Research, 1*(1), 010104-1–010104-11.

Kohl, P.B., Rosengrant, D., and Finkelstein, N.D. (2007). Strongly and weakly directed approaches to teaching multiple representation use in physics. *Physical Review Special Topics—Physics Education Research, 3*(1), 010108-1–010108-10.

Kolb, D.A. (1984). *Experiential learning.* Englewood Cliffs, NJ: Prentice Hall.

Kotovsky, K., Hayes, J.R., and Simon, H.A. (1985). Why are some problems hard? Evidence from tower of Hanoi. *Cognitive Psychology, 17*(2), 248-294.

Kowalski, P., and Taylor, A.K. (2009). The effect of refuting misconceptions in the introductory psychology class. *Teaching of Psychology, 36*(3), 153-159.

Kozhevnikov, M., and Thornton, R. (2006). Real-time data display, spatial visualization ability, and learning force and motion concepts. *Journal of Science Education and Technology, 15*(1), 111-132.

Kozhevnikov, M., Hegarty, M., and Mayer, R.E. (2002). Revising the visualizer-verbalizer dimension: Evidence for two types of visualizers. *Cognition and Instruction, 20*(1), 47-77.

Kozma, R.B., and Russell, J. (1997). Multimedia and understanding: Expert and novice responses to different representations of chemical phenomena. *Journal of Research in Science Teaching, 34*(9), 949-968.

Kriz, S., and Hegarty, M. (2007). Top-down and bottom-up influences on learning from animations. *International Journal of Human Computer Studies, 65*(11), 911-930.

Lajoie, S.P. (2003). Transitions and trajectories for studies of expertise. *Educational Researcher, 32*(8), 21-25.

Landy, D., and Goldstone, R.L. (2007). How abstract is symbolic thought? *Journal of Experimental Psychology: Learning Memory and Cognition, 33*(4), 720-733.

Larkin, J.H. (1979). Processing information for effective problem solving. *Engineering Education, 70*(3), 285-288.

Larkin, J.H. (1981a). Cognition of learning physics. *American Journal of Physics, 49*(6), 534-541.

Larkin, J.H. (1981b). Enriching formal knowledge: A model for learning to solve textbook physics problems. In J.R. Anderson (Ed.), *Cognitive skills and their acquisition* (pp. 311-334). Hillsdale, NJ: Lawrence Erlbaum Associates.

Larkin, J.H. (1983). Role of problem representation in physics. In D. Gentner and A. Stevens (Eds.), *Mental models* (pp. 75-98). Hillsdale, NJ: Lawrence Erlbaum Associates.

Larkin, J.H., and Simon, H.A. (1987). Why a diagram is (sometimes) worth ten thousand words. *Cognitive Science, 11*(1), 65-100.

Larkin, J.H., McDermott, J., Simon, D.P., and Simon, H.A. (1980). Models of competence in solving physics problems. *Cognitive Science, 4*(4), 317-345.

Lave, J., and Wenger, E. (1991). *Situated learning: Legitimate peripheral participation.* Cambridge, MA: Cambridge University Press.

Lewis, S.E., and Lewis, J.E. (2005). Departing from lectures: An evaluation of a peer-led guided inquiry alternative. *Journal of Chemical Education, 82*(1), 135-139.

Libarkin, J., and Brick, C. (2002). Research methodologies in science education: Visualization and the geosciences. *Journal of Geoscience Education, 50*, 449-455.

Libarkin, J.C., Kurdziel, J.P., and Anderson, S.W. (2007). College student conceptions of geological time and the disconnect between ordering and scale. *Journal of Geoscience Education, 55*(5), 413-422.

Liben, L.S. (1997). Children's understanding of spatial representations of place: Mapping the methodological landscape. In N. Foreman and R. Gillett (Eds.), *A handbook of spatial research paradigms and methodologies, Vol. 1: Spatial cognition in the child and adult* (pp. 41-82). East Sussex, UK: The Psychology Press (Taylor and Francis Group).

Liben, L.S., Kastens, K.A., and Christensen, A. (2011). Spatial foundations of science education: The illustrative case of instruction on introductory geological concepts. *Cognition and Instruction, 29*(1), 1-43.

Linenberger, K.J., and Bretz, S.L. (2012). Generating cognitive dissonance in student interviews through multiple representations. *Chemistry Education Research and Practice, 13*(3), 172-178.

Litzinger, T., Van Meter, P., Kapli, N., Zappe, S., and Toto, R. (2010). Translating education research into practice within an engineering education center: Two examples related to problem solving. *International Journal of Engineering Education, 26*(4), 860-868.

Lynch, M. (1990). The externalized retina: Selection and mathematization in the visual documentation of objects in the life sciences. In M. Lynch and S. Woolgar (Eds.), *Representation in scientific practice* (pp. 153-186). Cambridge, MA: MIT Press.

Macdonald, R.H., Manduca, C.A., Mogk, D.W., and Tewksbury, B.J. (2005). Teaching methods in undergraduate geoscience courses: Results of the 2004 On the Cutting Edge survey of U.S. faculty. *Journal of Geoscience Education, 53*(3), 237-252.

Manduca, C., and Kastens, K.A. (2012). Geoscience and geoscientists: Uniquely equipped to study the Earth. *Geological Society of America Special Papers, 486*, 1-12.

Markman, A.B. (1999). *Knowledge representation.* Mahwah, NJ: Lawrence Erlbaum Associates.

Marshall, S.P. (1995). *Schemas in problem solving.* New York: Cambridge University Press.

Martin, S.A., and Bassok, M. (2005). Effects of semantic cues on mathematical modeling: Evidence from word-problem solving and equation construction tasks. *Memory and Cognition, 33*(3), 471-478.

Martinez, M.E. (2010). *Learning and cognition: The design of the mind.* Upper Saddle River, NJ: Merrill.

Maskall, J., and Stokes, A. (2008). *Designing effective fieldwork for the environmental and natural sciences.* Plymouth, UK: Higher Education Academy Subject Centre for Geography, Earth and Environmental Sciences.

May, C.L., Eaton, L.S., and Whitmeyer, S.J. (2009). Integrating student-led research in fluvial geomorphology into traditional field courses: A case study from James Madison University's field course in Ireland. *Geological Society of America Special Papers, 461,* 195-204.

McClary, L., and Talanquer, V. (2010). College chemistry students' mental models of acids and acid strength. *Journal of Research in Science Teaching, 48*(1), 396-413.

McCracken, W.M., and Newstetter, W. (2001). *Text to diagram to symbol: Representational transformations in problem-solving.* Proceedings of the American Society of Engineering Education/IEEE Frontiers in Education Conference, Reno, NV.

McDermott, L.C., Rosenquist, M.L., and van Zee, E.H. (1987). Student difficulties in connecting graphs and physics: Examples from kinematics. *American Journal of Physics 55*(6), 503-513.

McKeithen, K.B., Reitman, J.S., Rueter, H.H., and Hirtle, S.C. (1981). Knowledge organization and skill differences in computer programmers. *Cognitive Psychology, 13*(3), 307-325.

McKim, R.H. (1980). *Thinking visually: A strategy manual for problem solving.* Belmont, CA: Lifetime Learning.

McLean, P., Johnson, C., Rogers, R., Daniels, L., Reber, J., Slator, B.M., Terpstra, J., and White, A. (2005). Molecular and cellular biology animations: Development and impact on student learning. *Cell Biology Education, 4,* 169-179.

Meltzer, D.E. (2005). Relation between students' problem-solving performance and representational format. *American Journal of Physics, 73*(5), 463-478.

Mestre, J.P. (2002). Probing adults' conceptual understanding and transfer of learning via problem posing. *Journal of Applied Developmental Psychology, 23*(1), 9-50.

Mestre, J.P., and Ross, B.H. (Eds.). (2011). *The psychology of learning and motivation* (vol. 55). San Diego, CA: Elsevier.

Meyer, J.H.F., and Land, R. (2005). Threshold concepts and troublesome knowledge (2): Epistemological considerations and a conceptual framework for teaching and learning. *Higher Education, 49*(3), 373-388.

Mogk, D., and Goodwin, C. (2012). Learning in the field: Synthesis of research on thinking and learning in the geosciences. *Geological Society of America Special Papers, 486,* 131-163.

Murphy, R., Ormand, C., Goodwin, L., Shipley, T., Manduca, C., and Tikoff, B. (2011, October). *Improving students' visuo-penetrative thinking skills through brief, weekly practice.* Paper presented at the Geological Society of America Annual Meeting, Minneapolis, MN.

Nachshon, I. (1985). Directional preferences in perception of visual stimuli. *International Journal of Behavioral Neuroscience, 25,* 161-174.

Nakhleh, M.B., and Mitchell, R.C. (1993). Concept learning versus problem solving: There is a difference. *Journal of Chemical Education, 70*(3), 190-192.

Niaz, M. (1989). Dimensional analysis: A neo-piagetian evaluation of m-demand of chemistry problems. *Research in Science & Technological Education, 7*(2), 153-170.

Nicoll, G. (2003). A qualitative investigation of undergraduate chemistry students' macroscopic interpretations of the submicroscopic structures of molecules. *Journal of Chemical Education, 80,* 205-213.

Novick, L.R. (1988). Analogical transfer, problem similarity, and expertise. *Journal of Experimental Psychology: Learning, Memory, and Cognition, 14*(3), 510-520.

Novick, L.R. (2001). Spatial diagrams: Key instruments in the toolbox for thought. In D.L. Medin (Ed.), *The psychology of learning and motivation* (vol. 40, pp. 279-325). San Diego, CA: Academic Press.

Novick, L.R., and Bassok, M. (2005). Problem solving. In K.J. Holyoak and R.G. Morrison (Eds.), *The Cambridge handbook of thinking and reasoning* (pp. 321-349). New York: Cambridge University Press.

Novick, L.R., and Catley, K.M. (2007). Understanding phylogenies in biology: The influence of a gestalt perceptual principle. *Journal of Experimental Psychology: Applied, 13*(4), 197-223.

Novick, L.R., and Catley, K.M. (in press). Reasoning about evolution's grand patterns: College students' understanding of the tree of life. *American Educational Research Journal.*

Novick, L.R., and Sherman, S.J. (2008). The effects of superficial and structural information on online problem solving for good versus poor anagram solvers. *Quarterly Journal of Experimental Psychology, 61*(7), 1098-1120.

Novick, L.R., Catley, K.M., and Funk, D.J. (2010). Characters are key: The effect of synapomorphies on cladogram comprehension. *Evolution: Education and Outreach, 3*, 539-547.

Novick, L.R., Catley, K.M., and Funk, D.J. (in press). Inference is bliss: Using evolutionary relationship to guide categorical inferences. *Cognitive Science.*

Novick, L.R., Catley, K.M., and Schreiber, E.G. (2010). *Understanding evolutionary history: An introduction to tree thinking (version 3).* Unpublished instructional booklet, Department of Psychology and Human Development. Nashville, TN: Vanderbilt University.

Novick, L.R., Stull, A.T., and Catley, K.M. (in press). Reading Phylogenetic trees: Effects of tree orientation and text processing on comprehension. *BioScience.*

Nurrenbern, S.C., and Pickering, M. (1987). Concept-learning versus problem-solving: Is there a difference? *Journal of Chemical Education, 64*(6), 508-510.

O'Day, D.H. (2007). The value of animations in biology teaching: A study of long-term memory retention. *CBE-Life Sciences Education, 6*(3), 217-223.

Ogilvie, C.A. (2009). Changes in students' problem-solving strategies in a course that includes context-rich, multifaceted problems. *Physical Review Special Topics—Physics Education Research, 5*(2).

Orion, N., Ben-Chaim, D., and Kali, Y. (1997). Relationship between earth-science education and spatial visualization. *Journal of Geoscience Education, 45*(2), 129-132.

Orion, N., Hofstein, A., Tamir, P., and Giddings, G.J. (1997). Development and validation of an instrument for assessing the learning environment of outdoor science activities. *Science Education, 81*(2), 161-171.

Orton, A. (1983). Students' understanding of integration. *Educational Studies in Mathematics, 14*(1), 1-18.

Owen, E., and Sweller, J. (1985). What do students learn while solving mathematics problems? *Journal of Educational Psychology, 77*, 272-284.

Ozdemir, G., Piburn, M., and Sharp, T. (2004). *Exploring the use of visualization in a college mineralogy course.* Paper presented at the National Association for Research in Science Teaching, Vancouver, BC.

Petcovic, H.L., and Libarkin, J.C. (2007). Research in science education: The expert-novice continuum. *Journal of Geoscience Education, 55*(4), 333-339.

Petcovic, H.L., Libarkin, J.C., and Baker, K.M. (2009). An empirical methodology for investigating geocognition in the field. *Journal of Geoscience Education, 57*(4), 316-328.

Piaget, J. (1978). *Success and understanding.* Cambridge, MA: Harvard University Press.

Piburn, M.D., Reynolds, S.J., McAuliffe, C., Leedy, D.E., and Johnson, J.K. (2005). The role of visualization in learning from computer-based images. *International Journal of Science Education, 27*(5), 513-527.

Piburn, M.D., van der Hoeven Kraft, K., and Pacheco, H. (2011). *A new century for geoscience education research*. Paper presented at the Second Committee Meeting on the Status, Contributions, and Future Directions of Discipline-Based Education Research. Available: http://www7.nationalacademies.org/bose/DBER_Piburn_October_Paper.pdf.

Pólya, G. (1945). *How to solve it, a new aspect of mathematical method*. Princeton, NJ: Princeton University Press.

Posner, M.I. (1973). *Cognition: An introduction*. Glenview, IL: Scott, Foresman.

Potter, N. Jr., Niemitz, J.W., and Sak, P.B. (2009). Long-term field-based studies in geoscience teaching. *Geological Society of America Special Papers, 461*, 185-194.

Previc, F.H. (1998). The neuropsychology of 3-D space. *Psychological Bulletin, 124*, 123-164.

Pribyl, J.R., and Bodner, G.M. (1987). Spatial ability and its role in organic-chemistry: A study of four organic courses. *Journal of Research in Science Teaching, 24*(3), 229-240.

Reif, F. (1995). *Understanding basic mechanics*. New York: John Wiley & Sons.

Reif, F., and Heller, J. (1982). Knowledge structure and problem solving in physics. *Educational Psychologist, 17*(2), 102-127.

Reif, F., and Scott, L.A. (1999). Teaching scientific thinking skills: Students and computers coaching each other. *American Journal of Physics, 67*(9), 819-831.

Reitman, W.R. (1965). *Cognition and thought, an information-processing approach*. New York: John Wiley & Sons.

Resnick, L.B. (1991). Shared cognition: Thinking as social practice. In L. Resnick, J. Levine, and S. Teasley (Eds.), *Perspectives on socially shared cognition* (pp. 127-149). Hyattsville, MD: American Psychological Association.

Riggs, E.M., Balliet, R., and Lieder, C. (2009). Using GPS tracking to understand problem solving during geologic field examinations. *Geological Society of America Special Papers, 461*.

Riggs, E.M., Lieder, C.C., and Balliet, R. (2009). Geologic problem solving in the field: Analysis of field navigation and mapping by advanced undergraduates. *Journal of Geoscience Education, 57*(1), 48-63.

Rosengrant, D., Van Heuvelen, A., and Etkina, E. (2005). Case study: Students' use of multiple representations in problem solving. *AIP Conference Proceedings, 818*, 49-52.

Rosengrant, D., Etkina, E., and Van Heuvelen, A. (2007). An overview of recent research on multiple representations. *AIP Conference Proceedings, 883*, 149-152.

Sadler, T.D., Burgin, S., McKinney, L., and Ponjuan, L. (2010). Learning science through research apprenticeships: A critical review of the literature. *Journal of Research in Science Teaching, 47*(3), 235-256.

Sandi-Urena, S., and Cooper, M.M. (2009). Design and validation of an inventory to assess metacognitive skillfulness in chemistry problem solving. *Journal of Chemical Education, 86*, 240-245.

Sandi-Urena, S., Cooper, M.M., and Stevens, R.H. (2011). Enhancement of metacognition use and awareness by means of a collaborative intervention. *International Journal of Science Education, 33*(3), 323-340.

Sanger, M.J., and Badger, S.M. (2001). Using computer-based visualization strategies to improve students' understanding of molecular polarity and miscibility. *Journal of Chemical Education, 78*(10), 1412.

Sankar, C., Varma, V., and Raju, P. (2008). Use of case studies in engineering education: Assessment of changes in cognitive skills. *Journal of Professional Issues in Engineering Education and Practice, 134*(3), 287-296.

Sawada, D., Piburn, M.D., Judson, E., Turley, J., Falconer, K., Benford, R., and Bloom, I. (2002). Measuring reform practices in science and mathematics classrooms: The reformed teaching observation protocol. *School Science and Mathematics, 102*, 245-253.

Sawrey, B.A. (1990). Concept-learning versus problem-solving—revisited. *Journal of Chemical Education, 67*(3), 253-254.

Schoenfeld, A.H., and Herrmann, D.J. (1982). Problem perception and knowledge structure in expert and novice mathematical problem solvers. *Journal of Experimental Psychology-Learning Memory and Cognition, 8*(5), 484-494.

Schönborn, K.J., and Anderson, T.R. (2009). A model of factors determining students' ability to interpret external representations in biochemistry. *International Journal of Science Education, 31*(2), 193-232.

Schwartz, D.L., and Black, J.B. (1996). Shuttling between depictive models and abstract rules: Induction and fallback. *Cognitive Science, 20*, 457-497.

Shea, D.L., Lubinski, D., and Benbow, C.P. (2001). Importance of assessing spatial ability in intellectually talented young adolescents: A 20-year longitudinal study. *Journal of Educational Psychology, 93*(3), 604-614.

Simon, D.P., and Simon, H.A. (1978). Individual differences in solving physics problems. In R.S. Siegler (Ed.), *Children's thinking: What develops?* (pp. 325-348). Hillsdale, NJ: Lawrence Erlbaum Associates.

Simon, H.A. (1978). Information-processing theory of human problem solving. In W.K. Estes (Ed.), *Handbook of learning and cognitive processes* (vol. 5, pp. 271-295). Hillsdale, NJ: Lawrence Erlbaum Associates.

Singh, K., Granville, M., and Dika, S. (2002). Mathematics and science achievement: Effects of motivation, interest, and academic engagement. *The Journal of Educational Research, 95*(6), 323-332.

Smith, A.D., Mestre, J.P., and Ross, B.H. (2010). Eye-gaze patterns as students study worked-out examples in mechanics. *Physical Review Special Topics—Physics Education Research, 6*(2).

Smith, M.U. (1992). Expertise and the organization of knowledge: Unexpected differences among genetic counselors, faculty, and students on problem categorization tasks. *Journal of Research in Science Teaching, 29*, 179-205.

Smith, M.U., and Good, R. (1984). Problem solving and classical genetics: Successful versus unsuccessful performance. *Journal of Research in Science Teaching, 21*, 895-912.

Sorby, S.A. (2009). Educational research in developing 3-D spatial skills for engineering students. *International Journal of Science Education, 31*(3), 459-480.

Stevens, R., and Palacio-Cayetano, J. (2003). Design and performance frameworks for constructing problem-solving simulation. *Cell Biology Education, 2*, 162-178.

Stieff, M. (2011). When is a molecule three-dimensional? A task-specific role for imagistic reasoning in advanced chemistry. *Science Education, 95*(2), 310-336.

Stieff, M., Hegarty, M., and Dixon, B.L. (2010). Alternative strategies for spatial reasoning with diagrams. In A. Goel, M. Jamnik, and N.H. Narayanan (Eds.), *Diagrammatic representation and inference (Proceedings of Diagrams 2010)*. Berlin, Germany: Springer-Verlag.

Svinicki, M. (2011). *Synthesis of the research on teaching and learning in engineering since the implementation of ABET engineering criteria 2000*. Paper presented at the Second Committee Meeting on the Status, Contributions, and Future Directions of Discipline-Based Education Research. Available: http://www7.nationalacademies.org/bose/DBER_Svinicki_October_Paper.pdf.

Sweller, J. (1988). Cognitive load during problem solving: Effects on learning. *Cognitive Science, 12*, 257-285.

Sweller, J., and Levine, M. (1982). Effects of goal specificity on means-ends analysis and learning. *Journal of Experimental Psychology: Learning, Memory, and Cognition, 8*, 463-474.

Sweller, J., Mawer, R., and Ward, M. (1983). Development of expertise in mathematical problem solving. *Journal of Experimental Psychology: General, 112*, 634-656.

Swenson, S., and Kastens, K.A. (2011). Student interpretation of a global elevation map: What it is, how it was made, and what it is useful for. *Geological Society of America Special Papers, 474*, 189-211.

Taagepera, M., and Noori, S. (2000). Mapping students' thinking patterns in learning organic chemistry by the use of knowledge space theory. *Journal of Chemical Education, 77*(9), 1224-1229.

Taatgen, N.A., and Anderson, J.R. (2008). ACT-R. In R. Sun (Ed.), *Constraints in cognitive architectures* (pp. 170-185). Cambridge, MA: Cambridge University Press.

Taber, K. (2009). Learning at the symbolic level. In J.K. Gilbert and D.F. Treagust (Eds.), *Multiple representations in chemical education* (pp. 75-108). Dordrecht, The Netherlands: Springer.

Taylor, J.L., Smith, K.M., van Stolk, A.P., and Spiegelman, G.B. (2010). Using invention to change how students tackle problems. *CBE-Life Sciences Education, 9*, 504-512.

Thagard, P. (2005). *Mind: Introduction to cognitive science* (2nd ed.). Cambridge, MA: MIT Press.

Tien, L.T., Teichert, M.A., and Rickey, D. (2007). Effectiveness of a MORE laboratory module in prompting students to revise their molecular-level ideas about solutions. *Journal of Chemical Education, 84*(1), 175-181.

Tingle, J.B., and Good, R.G. (1990). Effects of cooperative grouping on stoichiometric problem solving in high school chemistry. *Journal of Research in Science Teaching, 27*(7), 671-683.

Titus, S., and Horsman, E. (2009). Characterizing and improving spatial visualization skills. *Journal of Geoscience Education, 57*(4), 242-254.

Tversky, B., Zacks, J., Lee, P.U., and Heiser, J. (2000). Lines, blobs, crosses and arrows. In M. Anderson, P. Cheng, and V. Haarslev (Eds.), *Theory and application of diagrams* (pp. 221-230). Berlin, Heidelberg: Springer-Verlag.

Tversky, B., Morrison, J.B., and Betrancourt, M. (2002). Animation: Can it facilitate? *International Journal of Human-Computer Studies, 57*, 247-262.

Van Heuvelen, A. (1991). Learning to think like a physicist: A review of research-based instructional strategies. *American Journal of Physics, 59*(10), 891-897.

Van Heuvelen, A. (1995). Experiment problems for mechanics. *The Physics Teacher, 33*, 176-180.

Van Heuvelen, A., and Etkina, E. (2006). *The physics active learning guide*. San Francisco, CA: Pearson, Addison Wesley.

Van Heuvelen, A., and Maloney, D. (1999). Playing physics jeopardy. *American Journal of Physics, 67*, 252-256.

Van Heuvelen, A., and Zou, X. (2001). Multiple representations of work and energy processes. *American Journal of Physics, 69*(2), 184-194.

Voss, J.F., Greene, T.R., Post, T., and Penner, B.C. (1983). Problem-solving skills in the social sciences. In G. Bower (Ed.), *The psychology of learning and motivation* (pp. 165-213). New York: Academic Press.

Wai, J., Lubinski, D., and Benbow, C.P. (2009). Spatial ability for STEM domains: Aligning over 50 years of cumulative psychological knowledge solidifies its importance. *Journal of Educational Psychology, 101*, 817-835.

Ward, M., and Sweller, J. (1990). Structuring effective worked examples. *Cognition and Instruction, 7*(1), 1-39.

Webb, R.M., Lubinski, D., and Benbow, C.P. (2007). Spatial ability: A neglected dimension in talent searches for intellectually precocious youth. *Journal of Educational Psychology, 99*, 397-420.

Wheatley, G.H. (1984). *Problem solving in school mathematics*. MEPS Technical Report 84.01. West Lafayette, IN: Purdue University School Mathematics and Science Center.

Whitson, L., Bretz, S.L., and Towns, M.H. (2008). Characterizing the level of inquiry in the undergraduate laboratory. *Journal of College Science Teaching, 37*(7), 52-58.

Winn, W. (1989). The design and use of instructional graphics. In H. Mandl and J.R. Levin (Eds.), *Knowledge acquisition from text and pictures* (pp. 125-144). Amsterdam: Elsevier.

Wu, H., and Shah, P. (2004). Exploring visuospatial thinking in chemistry learning. *Science Education, 88*(3), 465-492.

Yerushalmi, E., Henderson, C., Heller, K., Heller, P., and Kuo, V. (2007). Physics faculty beliefs and values about the teaching and learning of problem solving. I: Mapping the common core. *Physical Review Special Topics—Physics Education Research, 3*(2), 020109-1-020109-31.

Yildirim, T., Shuman, L., and Besterfield-Sacre, M. (2010). Model-eliciting activities: Assessing engineering student problem solving and skill integration processes. *International Journal of Engineering Education, 26*(4), 831-845.

# CHAPTER 6

Adams, J.P., and Slater, T.F. (1998). *Mysteries of the sky*. Dubuque, IA: Kendall/Hunt.

Adams, J.P., and Slater, T.F. (2002). Learning through sharing: Supplementing the astronomy lecture with collaborative-learning group activities. *Journal of College Science Teaching, 31*(6), 384-387.

Adams, J.P., Slater, T.F., Lindell-Adrian, R., Brissenden, G., and Wallace, J. (2002). Observations of student behavior in collaborative learning groups. *Astronomy Education Review, 1*(1), 25-32.

Alexander, W.R. (2005). Assessment of teaching approaches in an introductory astronomy college classroom. *Astronomy Education Review, 3*(2), 178-186.

Allen, D., and Tanner, K. (2009). *Transformations: Approaches to college science teaching* (1st ed.). New York: W.H. Freeman.

Anderson, W.L., Mitchell, S.M., and Osgood, M.P. (2008). Gauging the gaps in student problem-solving skills: Assessment of individual and group use of problem-solving strategies using online discussions. *CBE-Life Sciences Education, 7*, 254-262.

Anderson, W.L., Sensibaugh, C.A., Osgood, M.P., and Mitchell, S.M. (2011). What really matters: Assessing individual problem-solving performance in the context of biological sciences. *International Journal for the Scholarship of Teaching and Learning, 5*, 1.

Armstrong, N., Chang, S.M., and Brickman, M. (2007). Cooperative learning in industrial-sized biology classes. *CBE-Life Sciences Education, 6*(2), 163-171.

Association of American Universities. (2011). *Five-year initiative for improving undergraduate STEM education: Discussion draft*. Washington, DC: Association of American Universities.

Bailey, J.M. (2011). *Astronomy education research: Developmental history of the field and summary of the literature*. Paper presented at the Second Committee Meeting on the Status, Contributions, and Future Directions of Discipline-Based Education Research. Available: http://www7.nationalacademies.org/bose/DBER_Bailey_Commissioned%20Paper.pdf.

Bailey, J.M., and Nagamine, K. (2009). *Results from a case study investigation on the adoption of learner-centered strategies in introductory astronomy.* Paper presented at the European Association for Research on Learning and Instruction Annual Meeting, Amsterdam, The Netherlands.

Barlow, A.E.L., and Villarejo, M. (2004). Making a difference for minorities: Evaluation of an educational enrichment program. *Journal of Research in Science Teaching, 41*(9), 861-881.

Beatty, I.D., Gerace, W.J., Leonard, W.J., and Dufresne, R.J. (2006). Designing effective questions for classroom response system teaching. *American Journal of Physics, 74*(1), 31-39.

Beichner, R.J. (1990). The effect of simultaneous motion presentation and graph generation in a kinematics lab. *Journal of Research in Science Teaching, 27*(8), 803-815.

Beichner, R.J. (1996). The impact of video motion analysis on kinematics graph interpretation skills. *American Journal of Physics, 64*(10), 1272-1277.

Beichner, R.J., Saul, J.M., Abbott, D.S., Morse, J.J., Deardorff, D.L., Allain, R.J., Bonham, S.W., Dancy, M.H., and Risley, J.S. (2007). The student-centered activities for large enrollment undergraduate programs (SCALE-UP) project. *Reviews in PER Volume 1: Research-based reform of university physics.* College Park, MD: American Association of Physics Teachers.

Bowen, C.W., and Phelps, A.J. (1997). Demonstration-based cooperative testing in general chemistry: A broader assessment-of-learning technique. *Journal of Chemical Education, 74*(6), 715-719.

Boyle A.P., Maguire S., et al. (2004). Is fieldwork good? An analysis of the student view. *Geological Society of America Abstracts with Programs, 36*(5), 156.

Brasell, H. (1987). The effect of real-time laboratory graphing on learning graphic representations of distance and velocity. *Journal of Research in Science Teaching, 24*(4), 385-395.

Brewe, E. (2008). Modeling theory applied: Modeling instruction in introductory physics. *American Journal of Physics, 76*(12), 1155-1160.

Brewe, E., Kramer, L., and O'Brien, G. (2009). Modeling instruction: Positive attitudinal shifts in introductory physics measured with CLASS. *Physical Review Special Topics—Physics Education Research, 5*(1), 013102-1–013102-5.

Brickman, P., Gormally, C., Armstrong, N., and Hallar, B. (2009). Effects of inquiry-based learning on students' science literacy skills and confidence. *International Journal for the Scholarship of Teaching and Learning, 3*(2).

Brogt, E., Sabers, D., Prather, E.E., Deming, G.L., Hufnagel, B., and Slater, T.F. (2007). Analysis of the astronomy diagnostic test. *Astronomy Education Review, 6*(1), 25-42.

Brown, J.S., Collins, A., and Duguid, P. (1989). Situated cognition and the culture of learning. *Educational Researcher, 18*(1), 32-42.

Bruck, L.B., Towns, M., and Bretz, S.L. (2010). Faculty perspectives of undergraduate chemistry laboratory: Goals and obstacles to success. *Journal of Chemical Education, 87*(12), 1416-1424.

Brungardt, J.B., and Zollman, D. (1995). Influence of interactive videodisc instruction using simultaneous-time analysis on kinematics graphing skills of high school physics students. *Journal of Research in Science Teaching, 32*(8), 855-869.

Butler, R. (2008). *Teaching geoscience through fieldwork, GEES subject centre learning and teaching guide.* Plymouth, UK: Higher Education Academy Subject Centre for Geography, Earth and Environmental Science. Available: http://www.gees.ac.uk/pubs/guides/fw/fwgeosci.pdf.

Caldwell, J.E. (2007). Clickers in the large classroom: Current research and best-practice tips. *CBE-Life Sciences Education, 6*(1), 9-20.

Catrambone, R., and Holyoak, K.J. (1989). Overcoming contextual limitations on problem-solving transfer. *Journal of Experimental Psychology: Learning, Memory, and Cognition, 15*, 1147-1156.

Chasteen, S.V., and Pollock, S.J. (2008). Transforming upper-division electricity and magnetism. *Journal of Research in Science Teaching, 21*, 221-233.

Clary, R.M., and Wandersee, J.H. (2007). A mixed methods analysis of the effects of an integrative geobiological study of petrified wood in introductory college geology classrooms. *Journal of Research in Science Teaching, 44*(8), 1011-1035.

Clement, J. (2008). The role of explanatory models in teaching for conceptual change. In S. Vosniadou (Ed.), *International handbook of research on conceptual change* (pp. 417-452). New York: Routledge.

Cooper, M.M., Cox, C.T., Jr., Nammouz, M., Case, E., and Stevens, R.J. (2008). An assessment of the effect of collaborative groups on students' problem-solving strategies and abilities. *Journal of Chemical Education, 85*(6), 866-872.

Cortright, R.N., Collins, H.L., Rodenbaugh, D.W., and DiCarlo, S.E. (2003). Student retention of course content is improved by collaborative-group testing. *Advances in Physiology Education, 27*, 102-108.

Crouch, C.H., Fagen, A.P., Callan, J.P., and Mazur, E. (2004). Classroom demonstrations: Learning tools or entertainment? *American Journal of Physics, 72*(6), 835-838.

Cummings, K., Marx, J., Thornton, R., and Kuhl, D. (1999). Evaluating innovation in studio physics. *American Journal of Physics, 67*(7, Suppl. 1), S38-S44.

Derting, T.L., and Ebert-May, D. (2010). Learner-centered inquiry in undergraduate biology: Positive relationships with long-term student achievement. *CBE-Life Sciences Education, 9*(4), 462-472.

Dirks, C. (2011). *The current status and future direction of biology education research.* Paper presented at the Second Committee Meeting on the Status, Contributions, and Future Directions of Discipline-Based Education Research. Available: http://www7.nationalacademies.org/bose/DBER_Dirks_October_Paper.pdf.

Dirks, C., and Cunningham, M. (2006). Enhancing diversity in science: Is teaching science process skills the answer? *CBE-Life Sciences Education, 5*(3), 218-226.

Docktor, J.L., and Mestre, J.P. (2011). *A synthesis of discipline-based education research in physics.* Paper presented at the Second Committee Meeting on the Status, Contributions, and Future Directions of Discipline-Based Education Research. Available: http://www7.nationalacademies.org/bose/DBER_Docktor_October_Paper.pdf.

Domin, D.S. (1999). A review of laboratory instruction styles. *Journal of Chemical Education, 76*(4), 543-547.

Elby, A. (2001). Helping physics students learn how to learn. *American Journal of Physics, 69* (7, Suppl. 1), S54-S64.

Elkins, J.T., and Elkins, N.M.L. (2007). Teaching geology in the field: Significant geoscience concept gains in entirely field-based introductory geology courses. *Journal of Geoscience Education, 55*(2), 126-132.

Elliott, M.J., Stewart, K.K., and Lagowski, J.J. (2008). The role of the laboratory in chemistry instruction. *Journal of Chemical Education, 85*(1), 145-149.

Etkina, E., and Van Heuvelen, A. (2007). Investigative science learning environment: A science-process approach to learning physics. In E.F. Redish and P.J. Cooney (Eds.), *Research-based reform of university physics* (vol. 1). College Park, MD: American Association of Physics Teachers. Available: http://www.per-central.org/document/ServeFile.cfm?ID=4988.

Etkina, E., Gibbons, K., Holton, B.L., and Horton, G.K. (1999). Lessons learned: A case study of an integrated way of teaching introductory physics to at-risk students at Rutgers University. *American Journal of Physics, 67*(9), 810-818.

Etkina, E., Van Heuvelen, A., White-Brahmia, S., Brookes, D.T., Gentile, M., Murthy, S., Rosengrant, D., and Warren, A. (2006). Scientific abilities and their assessment. *Physical Review Special Topics—Physics Education Research*, 2(2), 020103-1–020103-15.

Etkina, E., Karelina, A., Ruibal-Villasenor, M., Rosengrant, D., Jordan, R., and Hmelo-Silver, C.E. (2010). Design and reflection help students develop scientific abilities: Learning in introductory physics laboratories. *Journal of the Learning Sciences*, 19(1), 54-98.

Fay, M.E., Grove, N.P., Towns, M.H., and Bretz, S.L. (2007). A rubric to characterize inquiry in the undergraduate chemistry laboratory. *Chemistry Education Research and Practice*, 8(2), 212-219.

Feisel, L.D., and Peterson, G.D. (2002). *A colloquy on learning objectives for engineering education laboratories*. Proceedings of the 2002 American Society for Engineering Education Annual Conference and Exposition. Washington, DC: American Society for Engineering Education.

Feisel, L.D., and Rosa, A.J. (2005). The role of the laboratory in undergraduate engineering education. *Journal of Engineering Education*, 94(1), 121-130.

Finkelstein, N.D., Adams, W.K., Keller, C.J., Kohl, P.B., Perkins, K.K., Podolefsky, N.S., Reid, S., and LeMaster, R. (2005). When learning about the real world is better done virtually: A study of substituting computer simulations for laboratory equipment. *Physical Review Special Topics—Physics Education Research*, 1(1), 010103-1–010103-8.

Formica, S.P., Easley, J.L., and Spraker, M.C. (2010). Transforming common-sense beliefs into Newtonian thinking through Just-in-Time Teaching. *Physical Review Special Topics—Physics Education Research*, 6(2), 020106-1–020106-7.

Gabel, D. (1999). Improving teaching and learning through chemistry education research: A look to the future. *Journal of Chemical Education*, 76(4), 548-554.

Gaffney, J.D.H., Richards, E., Kustusch, M.B., Ding, L., and Beichner, R.J. (2008). Scaling up education reform. *Journal of College Science Teaching*, 37(5), 48-53.

Gentner, D., and Colhoun, J. (2010). Analogical processes in human thinking and learning. In B. Glatzeder, V. Goel, and A. V. Muller (Eds.), *On thinking: Volume 2. Towards a theory of thinking* (pp. 35-48). Berlin, Germany: Springer-Verlag.

Gick, M.L., and Holyoak, K.J. (1983). Schema induction and analogical transfer. *Cognitive Psychology*, 15, 1-38.

Hake, R.R. (1998). Interactive-engagement versus traditional methods: A six-thousand-student survey of mechanics test data for introductory physics courses. *American Journal of Physics*, 66(1), 64-74.

Halme, D.G., Khodor, J., Mitchell, R., and Walker, G.C. (2006). A small-scale concept-based laboratory component: The best of both worlds. *CBE-Life Sciences Education*, 5(1), 41-51.

Hanauer, D.I., Jacobs-Sera, D., Pedulla, M.L., Cresawn, S.G., Hendrix, R.W., and Hatfull, G.F. (2006). Teaching scientific inquiry. *Science*, 314(5807), 1880-1881.

Handelsman, J., Miller, S., and Pfund, C. (2007). *Scientific teaching*. New York: Freeman.

Hart, C., Mulhall, P., Berry, A., Loughran, J., and Gunstone, R. (2000). What is the purpose of this experiment? Or can students learn something from doing experiments? *Journal of Research in Science Teaching*, 37(7), 655-675.

Hatfull, G.F., Jacobs-Sera, D, Lawrence, J.G., Pope, W.H., Russell, D.A., Ko, C.C., Weber, R.J., Patel, M.C., Germane, K.L., Edgar, R.H., Hoyte, N.N., Bowman, C.A., Tantoco, A.T., Paladin, E.C., Myers, M.S., Smith, A.L., Grace, M.S., Pham, T.T., O'Brien, M.B., Vogelsberger, A.M., Hryckowian, A.J., Wynalek, J.L., Donis-Keller, H., Bogel, M.W., Peebles, C.L., Cresawn, S.G., and Hendrix, R.W. (2010). Comparative genomic analysis of sixty mycobacteriophage genomes: Genome clustering, gene acquisition and gene size. *Journal of Molecular Biology*, 397(1), 119-143.

Hays, J.D., Pfirman, S., Blumenthal, M., Kastens, K., and Menke, W. (2000). Earth science instruction with digital data. *Computers and the Geosciences, 26,* 657-668.

Heller, K., and Heller, P. (2000). *Competent problem solver—calculus version.* New York: McGraw-Hill.

Heller, P., and Hollabaugh, M. (1992). Teaching problem-solving through cooperative grouping. Part 2: Designing problems and structuring groups. *American Journal of Physics, 60*(7), 637-644.

Heller, P., Keith, R., and Anderson, S. (1992). Teaching problem solving through cooperative grouping. Part 1: Group versus individual problem solving. *American Journal of Physics, 60*(7), 627-636.

Herrington, D.G., and Nakhleh, M.B. (2003). What defines effective chemistry laboratory instruction? Teaching assistant and student perspectives. *Journal of Chemical Education, 80*(10), 1197-1205.

Hoellwarth, C., Moelter, M.J., and Knight, R.D. (2005). A direct comparison of conceptual learning and problem-solving ability in traditional and studio style classrooms. *American Journal of Physics, 73*(5), 459-462.

Hofstein, A., and Lunetta, V.N. (2004). The laboratory in science education: Foundations for the twenty-first century. *Science Education, 88*(1), 28-54.

Hofstein, A., and Mamlok-Naaman, R. (2007). The laboratory in science education: The state of the art. *Chemistry Education: Research and Practice in Europe, 8*(2), 105-108.

Hollabaugh, M. (1995). *Physics problem-solving in cooperative learning groups.* Unpublished doctoral dissertation, University of Minnesota, Twin Cities. Available: http://groups.physics.umn.edu/physed/People/Hollabaugh%20Dissertation.pdf.

Hudgins, D.W., Prather, E.E., Grayson, D.J., and Smits, D.P. (2006). Effectiveness of collaborative ranking tasks on student understanding of key astronomy concepts. *Astronomy Education Review, 5*(1), 1-22.

Hufnagel, B. (2002). Development of the Astronomy Diagnostic Test. *Astronomy Education Review, 1*(1), 47-51.

Hufnagel, B., Slater, T.F., Deming, G.L., Adams, J.P., Adrien, R.L., Brick, C., and Zeilik, M. (2000). Pre-course results from the Astronomy Diagnostic Test. *Publications of the Astronomical Society of Australia, 17*(2), 152-155.

Huntoon, J.E., Bluth, G.J.S., and Kennedy, W.A. (2001). Measuring the effects of a research-based field experience on undergraduates and K-12 teachers. *Journal of Geoscience Education, 49*(3), 235-248.

Jacoby, L.L. (1978). On interpreting the effects of repetition: Solving a problem versus remembering a solution. *Journal of Verbal Learning and Verbal Behavior, 17,* 649-667.

Jalil, P.A. (2006). A procedural problem in laboratory teaching: Experiment and explain, or vice-versa? *Journal of Chemical Education, 83*(1), 159-163.

Johnson, D.W., Johnson, R.T., and Holubec, E. (1990). *Circles of learning: Cooperation in the classroom.* Edina, MN: Interaction Book.

Johnson, D.W., Johnson, R.T., and Smith, K.A. (1991). *Cooperative learning: Increasing college faculty instructional productivity.* ASHE-ERIC Higher Education Report No. 4. Washington, DC: George Washington University School of Education and Human Development.

Johnson, D.W., Johnson, R.T., and Smith K.A. (1998). Cooperative learning returns to college: What evidence is there that it works? *Change, 30,* 26-35.

Johnson, D.W., Johnson, R.T., and Smith, K.A. (2007). The state of cooperative learning in postsecondary and professional settings. *Educational Psychology Review, 19*(1), 15-29.

Johnson, D.W., Johnson, R.T., and Smith, K.A. (2011). Lecturing with informal cooperative learning groups. In J. Cooper and P. Robinson (Eds.), *Small group learning in higher education: Research and practice.* Oklahoma City, OK: New Forums Press.

Johnson, M.A., and Lawson, A.E. (1998). What are the relative effects of reasoning ability and prior knowledge on biology achievement in expository and inquiry classes? *Journal of Research in Science Teaching, 35*(1), 89-103.

Kanter, D.E., Smith, H.D., McKenna, A., Rieger, C., and Linsenmeier, R.A. (2003). *Inquiry-based laboratory instruction throws out the cookbook and improves learning.* Evanston, IL: Northwestern University.

Keller, C., Finkelstein, N.D., Perkins, K.K., Pollock, S.J., Turpen, C., and Dubson, M. (2007). Research-based practices for effective clicker use. *AIP Conference Proceedings, 951,* 128-131. Available: http://www.compadre.org/Repository/document/servefile.cfm?ID=9085&docid=2003.

Kirschner, P.A., and Meester, M.A.M. (1988). The laboratory in higher science education: Problems, premises and objectives. *Higher Education, 17*(1), 81-98.

Klein, G.A., Orasanu, J., Calderwood, R., and Zsambok, C.E. (Eds.). (1993). *Decision making in action: Models and methods.* Norwood, NJ: Ablex.

Knight, J.K., and Wood, W.B. (2005). Teaching more by lecturing less. *Cell Biology Education, 4*(4), 298-310.

Kortz, K.M., Smay, J.J., and Murray, D.P. (2008). Increasing learning in introductory geoscience courses using lecture tutorials. *Journal of Geoscience Education, 56*(3), 280-290.

Lasry, N. (2008). Clickers or flashcards: Is there really a difference? *Physics Teacher, 46*(4), 242-244.

Laws, P.W. (1991). Calculus-based physics without lectures. *Physics Today, 44*(12), 24-31.

Laws, P.W. (2004). A unit on oscillations: Determinism and chaos for introductory physics students. *American Journal of Physics, 72*(4).

Laws, P.W., Rosborough, P.J., and Poodry, F.J. (1999). Women's responses to an activity-based introductory physics program. *American Journal of Physics, 67*(7, Suppl. 1), S32-S37.

Lawson, A.E. (1988). A better way to teach biology. *American Biology Teacher, 50*(5), 266-274.

Lazarowitz, R., and Tamir, P. (1994). Research on using laboratory instruction in science. In D.L. Gabel (Ed.), *Handbook of research on science teaching and learning: A project of the National Science Teachers Association* (pp. 94-128). New York: Macmillan.

Len, P.M. (2007). Different reward structures to motivate student interaction with electronic response systems in astronomy. *Astronomy Education Review, 5*(2), 5-15.

Lewis, S.E., and Lewis, J.E. (2005). Departing from lectures: An evaluation of a peer-led guided inquiry alternative. *Journal of Chemical Education, 82*(1), 135-139.

Linneman, S., and Plake, T. (2006). Searching for the difference: A controlled test of just-in-time teaching for large-enrollment introductory geology courses. *Journal of Geoscience Education, 54*(1), 18-24.

Lipshitz, R., Klein, G., Orasanu, J., and Salas, E. (2001). Focus article: Taking stock of naturalistic decision making. *Journal of Behavioral Decision Making, 14*(5), 331-352.

Lopresto, M.C. (2010). Comparing a lecture with a tutorial in introductory astronomy. *Physics Education, 45*(2), 196-201.

Lopresto, M.C., and Murrell, S.R. (2009). Using the Star Properties Concept Inventory to compare instruction with lecture tutorials to traditional lectures. *Astronomy Education Review, 8*(1), 010105-1–010105-5.

Lord, T., and Orkwiszewski, T. (2006). Moving from didactic to inquiry-based instruction in a science laboratory. *American Biology Teacher, 68*(6), 342-345.

Luo, W. (2008). Just-in-Time-Teaching (JITT) improves students' performance in classes: Adaptation of JITT in four geography courses. *Journal of Geoscience Education, 56*(2), 166-171.

Lynch, D.R., Russell, J.S., Evans, J.C., and Sutterer, K.G. (2009). Beyond the cognitive: The affective domain, values, and the achievement of the vision. *Journal of Professional Issues in Engineering Education and Practice, 135*(1), 47-56.

Lyons, D.L. (2011). *Impact of backwards faded scaffolding approach to inquiry-based astronomy laboratory experiences on undergraduate non-science majors' views of scientific inquiry.* Ph.D. Dissertation, University of Wyoming.

MacArthur, J.R., and Jones, L.L. (2008). A review of literature reports of clickers applicable to college chemistry classrooms. *Chemistry Education Research and Practice, 9*(3), 187-195.

Malina, E.G., and Nakhleh, M.B. (2003). How students use scientific instruments to create understanding: CCD spectrophotometers. *Journal of Chemical Education, 80*(6), 691-698.

Marrs, K.A., and Novak, G. (2004). Just-in-time teaching in biology: Creating an active learner classroom using the internet. *Cell Biology Education, 3*(1), 049-061.

Marshall, S.P. (1995). *Schemas in problem solving.* New York: Cambridge University Press.

Matsui, J., Liu, R., and Kane, C.M. (2003). Evaluating a science diversity program at UC Berkeley: More questions than answers. *Cell Biology Education, 2*(2), 117-121.

Mayer, R.E. (2010). *Applying the science of learning.* Upper Saddle River, NJ: Pearson.

Mayer, R.E. (2011). *Applying the science of learning to undergraduate science education.* Paper presented at the Third Committee Meeting on Status, Contributions, and Future Directions of Discipline-Based Education Research. Available: http://www7.nationalacademies.org/bose/DBER_Mayer_December_Paper.pdf.

McConnell, D.A., Steer, D.N., and Owens, K.D. (2003). Assessment and active learning strategies for introductory geology courses. *Journal of Geoscience Education, 51*(2), 205-216.

McConnell, D.A., Steer, D.N., Owens, K., Borowski, W., Dick, J., Foos, A., Knott, J.R., Malone, M., McGrew, H., Van Horn, S., Greer, L., and Heaney, P.J. (2006). Using conceptests to assess and improve student conceptual understanding in introductory geoscience courses. *Journal of Geoscience Education, 54*(1), 61-68.

McDermott, L.C., and Shaffer, P.S. (2002). *Tutorials in introductory physics* (1st ed.). Upper Saddle River, NJ: Prentice Hall.

McDermott, L.C., and the Physics Education Group at the University of Washington. (1996a). *Physics by inquiry: An introduction to physics and the physical sciences* (vol. I). New York: John Wiley & Sons.

McDermott, L.C., and the Physics Education Group at the University of Washington. (1996b). *Physics by inquiry: An introduction to physics and the physical sciences* (vol. II). New York: John Wiley & Sons.

Mestre, J.P., Ross, B.H., Brookes, D.T., Smith, A.D., and Nokes, T.J. (2009). How cognitive science can promote conceptual understanding in physics classrooms. In I.M. Saleh and M.S. Khine (Eds.), *Fostering scientific habits of mind: Pedagogical knowledge and best practices in science education* (pp. 3-8). Rotterdam, The Netherlands: Sense.

Miller, L.S., Nakhleh, M.B., Nash, J.J., and Meyer, J.A. (2004). Students' attitudes toward and conceptual understanding of chemical instrumentation. *Journal of Chemical Education, 81*(12), 1801-1808.

Mogk, D., and Goodwin, C. (2012). Learning in the field: Synthesis of research on thinking and learning in the geosciences. *Geological Society of America Special Papers, 486,* 131-163.

Moravec, M., Williams, A., Aguilar-Roca, N., and O'Dowd, D.K. (2010). Learn before lecture: A strategy that improves learning outcomes in a large introductory biology class. *CBE-Life Sciences Education, 9*(4), 473-481.

National Research Council. (1999). *How people learn: Bridging research and practice.* M.S. Donovan, J.D. Bransford, and J.W. Pellegrino (Eds.). Committee on Learning Research and Educational Practice, Commission on Behavioral and Social Sciences and Education. Washington, DC: National Academy Press.

Nelson, K.G., Huysken, K., and Kilibarda, Z. (2010). Assessing the impact of geoscience laboratories on student learning: Who benefits from introductory labs? *Journal of Geoscience Education, 58*(1), 43-50.

Novak, G.M. (1999). *Just-in-time teaching: Blending active learning with web technology.* Upper Saddle River, NJ: Prentice Hall.

Novick, L.R., and Holyoak, K.J. (1991). Mathematical problem solving by analogy. *Journal of Experimental Psychology: Learning, Memory, and Cognition, 17*, 398-415.

Orgill, M., and Bodner, G.M. (2004). What research tells us about using analogies to teach chemistry. *Chemistry Education Research and Practice, 5*(1), 15-32.

Orgill, M., and Bodner, G.M. (2006). An analysis of the effectiveness of analogy use in college-level biochemistry textbooks. *Journal of Research in Science Teaching, 43*(10), 1040-1060.

Orgill, M., and Bodner, G.M. (2007). Locks and keys: An analysis of biochemistry students' use of analogies. *Biochemistry and Molecular Biology Education, 35*(4), 244-254.

Piaget, J. (1978). *Success and understanding.* Cambridge, MA: Harvard University Press.

Piburn, M., van der Hoeven Kraft, K., and Pacheco, H. (2011). *A new century for geoscience education research.* Paper presented at the Second Committee Meeting on the Status, Contributions, and Future Directions of Discipline-Based Education Research. Available: http://www7.nationalacademies.org/bose/DBER_Piburn_October_Paper.pdf.

Podolefsky, N.S., and Finkelstein, N.D. (2006). The use of analogy in learning physics: The role of representations. *Physics Review Special Topics—Physics Education Research, 2,* 020101. Available: http://prst-per.aps.org/abstract/PRSTPER/v2/i2/e020101.

Podolefsky, N.S., and Finkelstein, N.D. (2007a). Analogical scaffolding and the learning of abstract ideas in physics: Empirical studies. *Physics Review Special Topics—Physics Education Research, 3,* 020104. Available: http://prst-er.aps.org/abstract/PRSTPER/v3/i2/e020104.

Podolefsky, N.S., and Finkelstein, N.D. (2007b). Analogical scaffolding and the learning of abstract ideas in physics: An example from electromagnetic waves. *Physics Review Special Topics—Physics Education Research, 3,* 010109.

Prather, E.E., Slater, T.F., Adams, J., and Brissenden, B. (2007). *Lecture-tutorials for introductory astronomy* (2nd ed.). New York: Prentice Hall.

President's Council of Advisors on Science and Technology. (2012). *Engage to excel: Producing one million additional college graduates with degrees in science, technology, engineering, and mathematics.* Washington, DC: Executive Office of the President, President's Council of Advisors on Science and Technology.

Prince, M. (2004). Does active learning work? A review of the research. *Journal of Engineering Education, 93*(3), 223-231.

Pyle, E.J. (2009). The evaluation of field course experiences: A framework for development, improvement, and reporting. *Geological Society of America Special Papers, 461,* 341-356.

Rao, S.P., Collins, H.L., and DiCarlo, S.E. (2002). Collaborative testing enhances student learning. *American Journal of Physiology—Advances in Physiology Education, 26*(1-4), 37-41.

Rath, K.A., Peterfreund, A.R., Xenos, S.P., Bayliss, F., and Carnal, N. (2007). Supplemental instruction in Introductory Biology I: Enhancing the performance and retention of under-represented minority students. *CBE-Life Sciences Education, 6*(3), 203-216.

Redish, E.F., Steinberg, R.N., and Saul, J.M. (1998). Student expectations in introductory physics. *American Journal of Physics, 66,* 212-224.

Riggs, E.M., Balliet, R., and Lieder, C.C. (2009). Using GPS tracking to understand problem solving during geologic field examinations. *Geological Society of America Special Papers, 461.*

Riggs, E.M., Lieder, C.C., and Balliet, R. (2009). Geologic problem solving in the field: Analysis of field navigation and mapping by advanced undergraduates. *Journal of Geoscience Education*, *57*(1), 48-63.

Rissing, S.W., and Cogan, J.G. (2009). Can an inquiry approach improve college student learning in a teaching laboratory? *CBE-Life Sciences Education*, *8*(1), 55-61.

Ross, B.H., and Kennedy, P.T. (1990). Generalizing from the use of earlier examples in problem solving. *Journal of Experimental Psychology: Learning, Memory, and Cognition*, *16*, 42-55.

Rudd, J.A., Greenbowe, T.J., and Hand, B.M. (2007). Using the science writing heuristic to improve students' understanding of general equilibrium. *Journal of Chemical Education*, *84*(12), 2007-2011.

Ruiz-Primo, M.A., Briggs, D., Iverson, H., Talbot, R., and Shepard, L.A. (2011). Impact of undergraduate science course innovations on learning. *Science*, *331*(6022), 1269-1270.

Sandi-Urena, S., Cooper, M., Gatlin, T. and Bhattacharyya, G. (2011). Students' experience in a general chemistry cooperative problem-based laboratory. *Chemistry Education Research and Practice*, *12*, 434-442.

Sandi-Urena, S., Cooper, M., Gatlin, T., Bhattacharyya, G., and Stevens, R. (in press). Effect of cooperative problem-based lab instruction on metacognition and problem-solving skills. *Journal of Chemical Education*.

Schwartz, D.L., and Bransford, J.D. (1998). A time for telling. *Cognition and Instruction*, *16*(4), 475-522.

Seymour, E., and Hewitt, N. (1997). *Talking about leaving: Why undergraduates leave the sciences*. Boulder, CO: Westview Press.

Seymour, E., Hunter, A.B., Laursen, S.L., and Deantoni, T. (2004). Establishing the benefits of research experiences for undergraduates in the sciences: First findings from a three-year study. *Science Education*, *88*(4), 493-534.

Shaffer, C.D., Alvarez, C., Bailey, C., Barnard, D., Bhalla, S., Chandrasekaran, C., Chandrasekaran, V., Chung, H.M., Dorer, D.R., Du, C., Eckdahl, T.T., Poet, J.L., Frohlich, D., Goodman, A.L., Gosser, Y., Hauser, C., Hoopes, L.L., Johnson, D., Jones, C.J., Kaehler, M., Kokan, N., Kopp, O.R., Kuleck, G.A., McNeil, G., Moss, R., Myka, J.L., Nagengast, A., Morris, R., Overvoorde, P.J., Shoop, E., Parrish, S., Reed, K., Regisford, E.G., Revie, D., Rosenwald, A.G., Saville, K., Schroeder, S., Shaw, M., Skuse, G., Smith, C., Smith, M., Spana, E.P., Spratt, M., Stamm, J., Thompson, J.S., Wawersik, M., Wilson, B.A., Youngblom, J., Leung, W., Buhler, J., Mardis, E.R., Lopatto, D., and Elgin, S.C. (2010). The genomics education partnership: Successful integration of research into laboratory classes at a diverse group of undergraduate institutions. *CBE-Life Sciences Education*, *9*(1), 55-69.

Sibbernsen, K.J. (2010). *The impact of collaborative groups versus individuals in undergraduate inquiry-based astronomy laboratory learning experiences*. Ph.D. Dissertation, Capella University. Available: http://www.docstoc.com/docs/77505821/The-impact-of-collaborative-groups-versus-individuals-in-undergraduate-inquiry-based-astronomy-laboratory-learning-exercises.

Simmons, M.E., Wu, X.B., Knight, S.L., and Lopez, R.R. (2008). Assessing the influence of field- and GIS-based inquiry on student attitude and conceptual knowledge in an undergraduate ecology lab. *CBE-Life Sciences Education*, *7*(3), 338-345.

Skala, C., Slater, T.F., and Adams, J.P. (2000). Qualitative analysis of collaborative learning groups in large enrollment introductory astronomy. *Publications of the Astronomical Society of Australia*, *17*(2), 185-193.

Slater, S.J., Slater, T.F., and Shaner, A. (2008). Impact of backwards faded scaffolding in an astronomy course for pre-service elementary teachers based on inquiry. *Journal of Geoscience Education*, *56*(5), 408-416.

Slater, S.J., Slater, T.F., and Lyons, D. (2010). *Engaging in astronomical inquiry*. New York: W.H. Freeman.

Slavin, R.E., Hurley, E.A., and Chamberlain, A. (2003). Cooperative learning and achievement: Theory and practice. In W.M. Reynolds and G.E. Miller (Eds.), *Handbook of psychology* (vol. 7, pp. 177-198). New York: John Wiley & Sons.

Smith, K.A. (2000). Going deeper: Formal small-group learning in large classes. In J. Macgregor, J. Cooper, K. Smith, and P. Robinson (Eds.), *Strategies for energizing large classes: From small groups to learning communities: New directions for teaching and learning* (pp. 25-46). San Francisco, CA: Jossey-Bass.

Smith, K.A., Sheppard, S.D., Johnson, D.W., and Johnson, R.T. (2005). Pedagogies of engagement: Classroom-based practices. *Journal of Engineering Education, 94*(1), 87-100.

Smith, M.K., Wood, W.B., Adams, W.K., Wieman, C., Knight, J.K., Guild, N., and Su, T.T. (2009). Why peer discussion improves student performance on in-class concept questions. *Science, 323*(5910), 122-124.

Smith, M.K., Wood, W.B., Krauter, K., and Knight, J.K. (2011). Combining peer discussion with instructor explanation increases student learning from in-class concept questions. *CBE-Life Sciences Education, 10*(1), 55-63.

Sokoloff, D.R., and Thornton, R.K. (1997). Using interactive lecture demonstrations to create an active learning environment. *The Physics Teacher, 35*(6), 340-347.

Sokoloff, D.R., and Thornton R.K. (2004). *Interactive lecture demonstrations: Active learning in introductory physics*. New York: John Wiley & Sons.

Sorensen, C.M., Churukian, A.D., Maleki, S., and Zollman, D.A. (2006). The new studio format for instruction of introductory physics. *American Journal of Physics, 74*(12), 1077-1082.

Springer, L., Stanne, M.E., and Donovan, S.S. (1999). Effects of small-group learning on undergraduates in science, mathematics, engineering, and technology: A meta-analysis. *Review of Educational Research, 69*(1), 21-51.

Steinberg, R.N., Wittmann, M.C., and Redish, E.F. (1997). Mathematical tutorials in introductory physics. *AIP Conference Proceedings, 399*, 1075-1092.

Stelzer, T., Gladding, G., Mestre, J.P., and Brookes, D.T. (2009). Comparing the efficacy of multimedia modules with traditional textbooks for learning introductory physics content. *American Journal of Physics, 77*(2), 184-190.

Stokes, A., and Boyle, A. (2009). The undergraduate geoscience fieldwork experience: Influencing factors and implications for learning. *Geological Society of America Special Papers, 461*, 291-311.

Svinicki, M. (2011). *Synthesis of the research on teaching and learning in engineering since the implementation of ABET Engineering Criteria 2000*. Paper presented at the Second Committee Meeting on the Status, Contributions, and Future Directions of Discipline-Based Education Research. Available: http://www7.nationalacademies.org/bose/DBER_Svinicki_October_Paper.pdf.

Thacker, B., Eunsook, K., Kelvin, T., and Lea, S.M. (1994). Comparing problem-solving performance of physics students in inquiry-based and traditional introductory physics courses. *American Journal of Physics, 62*(7), 627-633.

Tien, L.T., Roth, V., and Kampmeier, J.A. (2002). Implementation of a peer-led team learning instructional approach in an undergraduate organic chemistry course. *Journal of Research in Science Teaching, 39*(7), 606-632.

Towns, M., and Kraft, A. (2011). *Review and synthesis of research in chemical education from 2000-2010*. Paper presented at the Second Committee Meeting on the Status, Contributions, and Future Directions of Discipline-Based Education Research. Available: http://www7.nationalacademies.org/bose/DBER_Towns_October_Paper.pdf.

Vygotsky, L.S. (1978). *Mind in society: The development of higher psychological processes.* Cambridge, MA: Harvard University Press.

White, R.T. (1996). The link between the laboratory and learning. *International Journal of Science Education, 18*(7), 761-774.

Whitmeyer, S., Mogk, D., and Pyle, E. (Eds). (2009). *Field geology education: Historical perspectives and modern approaches.* Boulder, CO: Geological Society of America.

Whitson, L., Raker, J., Bretz, S.L., and Towns, M.H. (2008). Characterizing the level of inquiry in the undergraduate laboratory. *Journal of College Science Teaching, 37*(7), 52-58.

Wieman, C., Perkins, K., Gilbert, S., Benay, F., Kennedy, S., Semsar, K., Knight, J., Shi, J., Smith, M., Kelly, T., Taylor, J., Yurk, H., Birol, G., Langdon, L., Pentecost, T., Stewart, J., Arthurs, L., Bair, A., Stempien, J., Gilley, B., Jones, F., Kennedy, B., Chasteen, S., and Simon, B. (2008). *Clicker resource guide: An instructor's guide to the effective use of personal response systems (clickers) in teaching.* Available: http://www.colorado.edu/sei/documents/clickeruse_guide0108.pdf.

Wittmann, M.C., Steinberg, R., and Redish, E. (2004). *Activity-based tutorials. Volume 1: Introductory physics.* Hoboken, NJ: John Wiley & Sons.

Wittmann, M.C., Steinberg, R., and Redish, E. (2005). *Activity-based tutorials. Volume 2: Modern physics.* Hoboken, NJ: John Wiley & Sons.

Wood, W.B. (2004). Clickers: A teaching gimmick that works. *Developmental Cell, 7*(6), 796-798.

Wood, W.B. (2009). Innovations in teaching undergraduate biology and why we need them. *Annual Review of Cell and Developmental Biology, 25,* 93-112.

Wright, R., and Boggs, J. (2002). Learning cell biology as a team: A project-based approach to upper-division cell biology. *Cell Biology Education, 1*(4), 145-153.

Yerushalmi, E., Henderson, C., Heller, K., Heller, P., and Kuo, V. (2007). Physics faculty beliefs and values about the teaching and learning of problem solving. I. Mapping the common core. *Physical Review Special Topics—Physics Education Research, 3*(2), 020109-1–020109-31.

Yuretich, R.F., Khan, S.A., Leckie, R.M., and Clement, J.J. (2001). Active-learning methods to improve student performance and scientific interest in a large introductory oceanography course. *Journal of Geoscience Education, 49*(2), 111-119.

Zsambok, C., and Klein, G. (1997). *Naturalistic decision making.* Mahwah, NJ: Lawrence Erlbaum Associates.

## CHAPTER 7

Adams, W.K., Perkins, K.K., Podolefsky, N.S., Dubson, M., Finkelstein, N.D., and Wieman, C.E. (2006). New instrument for measuring student beliefs about physics and learning physics: The Colorado Learning Attitudes about Science Survey. *Physical Review Special Topics—Physics Education Research, 2*(1).

Ainsworth, S., and Loizou, A.T. (2003). The effects of self-explaining when learning with text or diagrams. *Cognitive Science, 27,* 669-681.

Ambrose, S.A., Bridges, M.W., DiPietro, M., Lovett, M.C., Norman, M.K., and Mayer, R.E. (2010). *How learning works: Seven research-based principles for smart teaching.* San Francisco, CA: Jossey-Bass.

American Association of Physics Teachers. (1997). Goals of the introductory physics laboratory. *The Physics Teacher, 35,* 546-548.

Atman, C., Sheppard, S.D., Turns, J., Adams, R.S., Fleming, L.N., Stevens, R., Streveler, R.A., Smith, K.A., Miller, R.L., Leifer, L.J., Yasuhara, K., and Lund, D. (2010). *Enabling engineering student success: The final report for the Center for the Advancement of Engineering Education.* Technical Report CAEE-TR-10-02. San Rafael, CA: Morgan & Claypool.

Ausubel, D.P. (1968). *Educational psychology, a cognitive view.* New York: Holt, Rinehart and Winston.

Ausubel, D.P., Novak, J.D., and Hanesian, H. (1978). *Educational psychology: A cognitive view* (2d ed.). New York: Holt, Rinehart and Winston.

Bandura, A. (1986). *Social foundations of thought and action: A social cognitive theory.* Englewood Cliffs, NJ: Prentice-Hall.

Barbera, J., Ams, W.K., Wieman, C.E., and Perkins, K.K. (2008). Modifying and validating the Colorado Learning Attitudes about Science Survey for use in chemistry. *Journal of Chemical Education, 85,* 1435-1439.

Bassok, M. (1990). Transfer of domain-specific problem-solving procedures. *Journal of Experimental Psychology: Learning, Memory, and Cognition, 16*(3), 522-533.

Bassok, M. (2003). Analogical transfer in problem solving. In J.E. Davidson and R.J. Sternberg (Eds.), *Psychology of problem solving* (pp. 343-369). New York: Cambridge University Press.

Bassok, M., and Holyoak, K.J. (1989). Interdomain transfer between isomorphic topics in algebra and physics. *Journal of Experimental Psychology: Learning, Memory, and Cognition, 15*(1), 153-166.

Boyle, A., Maguire, S., Martin, A., Milsom, C., Nash, R., Rawlinson, S., Turner, A., Wurthmann, S., and Conchie, S. (2007). Fieldwork is good: The student perception and the affective domain. *Journal of Geography in Higher Education, 31*(2), 299-317.

Brandriet, A.R., Xu, X., Bretz, S.L., and Lewis, J.E. (2011). Diagnosing changes in attitude in first-year college chemistry students with a shortened version of Bauer's semantic differential. *Chemistry Education Research and Practice, 12*(2), 271-278.

Bransford, J.D., and Schwartz, D.L. (1999). Rethinking transfer: A simple proposal with multiple implications. *Review of Research in Education, 24,* 61-100.

Bretz, S.L. (2001). Novak's theory of education: Human constructivism and meaningful learning. *Journal of Chemical Education, 78*(8), 1107.

Brown, A.L. (1978). Knowing when, where, and how to remember: A problem of metacognition. In R. Glaser (Ed.), *Advances in instructional psychology* (vol. 2, pp. 77-165). Hillsdale, NJ: Lawrence Erlbaum Associates.

Brown, J.S., Collins, A., and Duguid, P. (1989). Situated cognition and the culture of learning. *Educational Researcher, 18*(1), 32-42.

Catrambone, R. (1998). The subgoal learning model: Creating better examples so that students can solve novel problems. *Journal of Experimental Psychology: General, 127,* 355-376.

Chi, M.T.H., and Bassok, M. (1989). Learning from examples via self-explanations. In L.B. Resnick (Ed.), *Knowing, learning, and instruction: Essays in honor of Robert Glaser* (pp. 251-282). Hillsdale, NJ: Lawrence Erlbaum Associates.

Chi, M.T.H., and VanLehn, K. (1991). The content of physics self-explanations. *Journal of the Learning Sciences, 1*(1), 69-106.

Chi, M.T.H., Bassok, M., Lewis, M.W., Reimann, P., and Glaser, R. (1989). Self-explanations: How students study and use examples in learning to solve problems. *Cognitive Science, 13*(2), 145-182.

Clark, I.F., and James, P.R. (2004). Using concept maps to plan an introductory structural geology course. *Journal of Geoscience Education, 52*(3), 224-230.

Coil, D., Wenderoth, M.P., Cunningham, M., and Dirks, C. (2010). Teaching the process of science: Faculty perceptions and an effective methodology. *CBE-Life Sciences Education, 9*(4), 524-535.

Collins, A., Brown, J.S., and Newman, S.E. (1989). Cognitive apprenticeship: Teaching the crafts of reading, writing, and mathematics. In L.B. Resnick (Ed.), *Knowing, learning, and instruction: Essays in honor of Robert Glaser* (pp. 453-494). Hillsdale, NJ: Lawrence Erlbaum Associates.

Collins, H., and Pinch, T. (1993). *The Golem: What everyone should know about science.* Cambridge, MA: Cambridge University Press.

Connor, C. (2009). Field glaciology and earth systems science: The Juneau Icefield Research Program (JIRP), 1946-2008. *Geological Society of America Special Papers, 461,* 173-184.

Cooper, M.M., and Sandi-Urena, S. (2009). Design and validation of an instrument to assess metacognitive skillfulness in chemistry problem solving. *Journal of Chemical Education, 86*(2), 240-245.

Cooper, M.M., Sandi-Urena, S., and Stevens, R. (2008). Reliable multimethod assessment of metacognition use in chemistry problem solving. *Chemistry Education Research and Practice, 9*(1), 18-24.

Cooper, M.M., Cox, C.T., Jr., Nammouz, M., Case, E., and Stevens, R.J. (2008). An assessment of the effect of collaborative groups on students' problem-solving strategies and abilities. *Journal of Chemical Education, 85*(6), 866-872.

Cox, A.J., and Junkin, W.F., III. (2002). Enhanced student learning in the introductory physics laboratory. *Physics Education, 37*(1), 37-44.

Craik, F.I.M, and Lockhart, R.S. (1972). Levels of processing: A framework for memory research. *Journal of Verbal Learning and Verbal Behavior, 11,* 671-684.

Crompton, J.L., and Sellar, C. (1981). Do outdoor education experiences contribute to positive development in the affective domain? *Journal of Environmental Education, 12*(4), 21-29.

Crouch, C.H., Fagen, A.P., Callan, J.P., and Mazur, E. (2004). Classroom demonstrations: Learning tools or entertainment? *American Journal of Physics, 72*(6), 835-838.

Cummings, K., Marx, J., Thornton, R., and Kuhl, D. (1999). Evaluating innovation in studio physics. *American Journal of Physics, 67*(7, Suppl. 1), S38-S44.

De Leone, C.J., and Gire, E. (2006). Is instructional emphasis on the use of non-mathematical representations worth the effort? *AIP Conference Proceedings, 818,* 45-48.

DeBoer, G.E. (1991). *A history of ideas in science education.* New York: Teachers College Press.

Dirks, C. (2011). *The current status and future direction of biology education research.* Paper presented at the Second Committee Meeting on the Status, Contributions, and Future Directions of Discipline-Based Education Research. Available: http://www7.national academies.org/bose/DBER_Dirks_October_Paper.pdf.

Dirks, C., and Cunningham, M. (2006). Enhancing diversity in science: Is teaching science process skills the answer? *CBE-Life Sciences Education, 5*(3), 218-226.

Dunbar, K. (1995). How scientists really reason: Scientific reasoning in real-world laboratories. In R.J. Sternberg and J.E. Davidson (Eds.), *Mechanisms of insight* (pp. 365-395). Cambridge, MA: MIT Press.

Elby, A. (2001). Helping physics students learn how to learn. *American Journal of Physics, 69*(7, Suppl. 1), S54-S64.

Elkins, J., Elkins, N.M.L., and Hemmings, S.N.J. (2008). GeoJourney: A nine-week introductory field program with an interdisciplinary approach to teaching geology, Native American cultures, and environmental studies. *Journal of College Science Teaching, 37*(3), 18-28.

Etkina, E., Karelina, A., Ruibal-Villasenor, M., Rosengrant, D., Jordan, R., and Hmelo-Silver, C.E. (2010). Design and reflection help students develop scientific abilities: Learning in introductory physics laboratories. *Journal of the Learning Sciences, 19*(1), 54-98.

Flavell, J.H. (1979). Metacognition and cognitive monitoring: A new area of cognitive-developmental inquiry. *American Psychologist, 34*(10), 906-911.

Flick, L., and Lederman, N.G. (Eds.). (2004). *Scientific inquiry and nature of science: Implications for teaching, learning, and teacher education.* The Netherlands: Kluwer Academic.

Fuller, I., Gaskin, S., and Scott, I. (2003). Student perceptions of geography and environmental science fieldwork in the light of restricted access to the field, caused by foot and mouth disease in the UK in 2001. *Journal of Geography In Higher Education, 27*(1), 79-102.

Fuller, I., Edmondson, S., France, D., Higgitt, D., and Ratinen, I. (2006). International perspectives on the effectiveness of geography fieldwork for learning. *Journal of Geography in Higher Education, 30*(1), 89-101.

Gawel, J.E., and Greengrove, C.L. (2005). Designing undergraduate research experiences for nontraditional student learning at sea. *Journal of Geoscience Education, 53,* 31-36.

Gilligan, M.R., Verity, P.G., Cook, C.B., Cook, S.B., Booth, M.G., and Frischer, M.E. (2007). *Building a diverse and innovative ocean workforce through collaboration and partnerships that integrate research and education: HBCUs and marine laboratories.* Available: http://www.skio.usg.edu/publications/downloads/pdfs/_pubs/GilliganEA07.pdf.

Glynn, S.M., and Koballa, T.R. (2006). Motivation to learn in college science. In J. Mintzes and W.H. Leonard (Eds.), *Handbook of college science teaching* (pp. 25-32). Arlington, VA: National Science Teachers Association Press.

Goertzen, R.M., Scherr, R.E., and Elby, A. (2009). Accounting for tutorial teaching assistants' buy-in to reform instruction. *Physical Review Special Topics—Physics Education Research, 5*(2), 020109.

Goertzen, R.M., Scherr, R.E., and Elby, A. (2010). Respecting tutorial instructors' beliefs and experiences: A case study of a physics teaching assistant. *Physical Review Special Topics—Physics Education Research, 6*(2), 020125-1–020125-11.

Gray, J.R. (2004). Integration of emotion and cognitive control. *Current Directions in Psychological Science, 13*(2), 46-48.

Gregerman, S.R. (2008). *The role of undergraduate research in student retention, academic engagement, and the pursuit of graduate education.* Paper presented at the National Research Council's Workshop on Linking Evidence to Promising Practices in STEM Undergraduate Education, Washington, DC. Available: http://www7.nationalacademies.org/bose/Gregerman_CommissionedPaper.pdf.

Grove, N., and Bretz, S.L. (2007). CHEMX: An instrument to assess students' cognitive expectations for learning chemistry. *Journal of Chemical Education, 84*(9), 1524-1529.

Gunter, M.E. (2004). The polarized light microscope: Should we teach the use of a 19th century instrument in the 21st century? *Journal of Geoscience Education, 52*(1), 34-44.

Halloun, I. (1997). Views about science and physics achievement: The VASS story. *AIP Conference Proceedings, 399,* 605-614.

Hammer, D. (1994). Epistemological beliefs in introductory physics. *Cognition and Instruction, 12*(2), 151-183.

Hammer, D. (1995). Epistemological considerations in teaching introductory physics. *Science Education, 79*(4), 393-413.

Hammer, D., Elby, A., Scherr, R., and Redish, E.F. (2005). Resources, framing, and transfer. In J. Mestre (Ed.), *Transfer of learning from a modern multidisciplinary perspective* (pp. 89-119). Greenwich, CT: Information Age.

Heller, P., Keith, R., and Anderson, S. (1992). Teaching problem solving through cooperative grouping. Part 1: Group versus individual problem solving. *American Journal of Physics, 60*(7), 627-636.

Henderson, C., and Dancy, M.H. (2007). Barriers to the use of research-based instructional strategies: The influence of both individual and situational characteristics. *Physical Review Special Topics—Physics Education Research, 3*(2), 020102-1–020102-14.

Henderson, C., Yerushalmi, E., Kuo, V.H., Heller, P., and Heller, K. (2004). Grading student problem solutions: The challenge of sending a consistent message. *American Journal of Physics, 72*(2), 164-169.

Henderson, C., Yerushalmi, E., Kuo, V.H., Heller, K., and Heller, P. (2007). Physics faculty beliefs and values about the teaching and learning of problem solving II: Procedures for measurement and analysis. *Physical Review Special Topics—Physics Education Research, 3*(2), 020110-1–020110-12.

Hilgard, E.R. (2006). The trilogy of mind: Cognition, affection and conation. *Journal of the History of the Behavioral Sciences, 16*(2), 107-117.

Holyoak, K.J., and Koh, K. (1987). Surface and structural similarity in analogical transfer. *Memory and Cognition, 15*(4), 332-340.

Hunter, A., Laursen, S., Seymour, E. (2007). Becoming a scientist: The role of undergraduate research in students' cognitive, personal and professional development. *Science Education, 91*(1), 36-74.

Huntoon, J.E., Bluth, G.J.S., and Kennedy, W.A. (2001). Measuring the effects of a research-based field experience on undergraduates and K-12 teachers. *Journal of Geoscience Education, 49*(3), 235-248.

Jarrett, O.S., and Burnley, P.C. (2010). Lessons on the role of fun/playfulness from a geology undergraduate summer research program. *Journal of Geoscience Education, 58*(2), 110-120.

Jee, B.D., Uttal, D.H., Gentner, D., Manduca, C., Shipley, T.F., Tikoff, B., Ormand, C.J., and Sageman, B. (2010). Commentary: Analogical thinking in geoscience education. *Journal of Geoscience Education, 58*(1), 2-13.

Kaminsky, J.A., and Sloutsky, V.M. (2012). Representation and transfer of abstract mathematical concepts in adolescence and young adulthood. In V.F. Reyna, S.B. Chapman, M.R. Dougherty, and J. Confrey (Eds.), *The adolescent brain: Learning, reasoning, and decision making* (pp. 67-93). Washington, DC: American Psychological Association.

Karabinos, P., Stoll, H.M., and Fox, W.T. (1992). Attracting students to science through field exercises in introductory geology courses. *Journal of Geological Education, 40*, 302-305.

Kardash, C.M. (2000). Evaluation of an undergraduate research experience: Perceptions of undergraduate interns and their faculty mentors. *Journal of Educational Psychology, 92*(1), 191-201.

Karelina, A., and Etkina, E. (2007). Acting like a physicist: Student approach study to experimental design. *Physical Review Special Topics—Physics Education Research, 3*.

Karpicke, J.D., Butler, A.C., and Roediger, H.L. (2009). Metacognitive strategies in student learning: Do students practice retrieval when they study on their own? *Memory, 17*, 471-479.

Karplus, R. (1977). Science teaching and the development of reasoning. *Journal of Research in Science Teaching, 14*, 169-175.

Kastens, K.A., Agrawal, S., and Liben, L.S. (2009). How students and field geologists reason in integrating spatial observations from outcrops to visualize a 3-D geological structure. *International Journal of Science Education, 31*(3), 365-393.

Kelly, R.J.L. (2007). Exploring how different features of animations of sodium chloride dissolution affect students' explanations. *Journal of Science Education and Technology, 16*(5), 413-429.

Kelly, R.M., and Jones, L.L. (2008). Investigating students' ability to transfer ideas learned from molecular animations of the dissolution process. *Journal of Chemical Education, 85*(2), 303-309.

Kempa, R.F., and Orion, N. (1996). Students' perception of co-operative learning in earth science fieldwork: *Research in Science and Technological Education, 14*(1), 33-41.

Kern, E.L., and Carpenter, J.R. (1984). Enhancement of student values, interests and attitudes in earth science through a field-oriented approach. *Journal of Geological Education, 32*(5), 299-305.

Kern, E.L., and Carpenter, J.R. (1986). Effect of field activities on student learning. *Journal of Geological Education, 34*(3), 180-183.

Kitchen, E., Bell, J.D., Reeve, S., Sudweeks, R.R., and Bradshaw, W.S. (2003). Teaching cell biology in the large-enrollment classroom: Methods to promote analytical thinking and assessment of their effectiveness. *Cell Biology Education, 2*(3), 180-194.

Kohl, P.B., and Finkelstein, N.D. (2005). Student representational competence and self-assessment when solving physics problems. *Physical Review Special Topics—Physics Education Research, 1*(1), 010104-1–010104-11.

Kost-Smith, L.E., Pollock, S.J., and Finkelstein, N.D. (2010). Gender disparities in second-semester college physics: The incremental effects of a smog of bias. *Physical Review Special Topics—Physics Education Research, 6*(2). Available: http://www.colorado.edu/physics/EducationIssues/papers/Kost_etal/gender_disparities.pdf.

Krathwohl, D.R., Bloom, B.S., and Masia, B.B. (1964). *Taxonomy of educational objectives: Handbook II: Affective domain.* New York: David McKay.

Lave, J., and Wenger, E. (1991). *Situated learning: Legitimate peripheral participation.* Cambridge, UK: Cambridge University Press.

Lawson, A.E. (1988). A better way to teach biology. *American Biology Teacher, 50*(5), 266-274.

Liben, L.S., Kastens, K.A., and Christensen, A. (2011). Spatial foundations of science education: The illustrative case of instruction on introductory geological concepts. *Cognition and Instruction, 29*(1), 1-43.

Lin, X., Schwartz, D.L., and Hatano, G. (2005). Toward teachers' adaptive metacognition. *Educational Psychologist, 40*(4), 245-255.

Locks, A.M., and Gregerman, S.R. (2008). Undergraduate research as an institutional retention strategy: The University of Michigan model. In R. Taraban and R.L. Blanton (Eds.), *Creating effective undergraduate research programs in science: The transformation from student to scientist* (pp. 11-32). New York: Teachers College Press.

Lopatto, D. (2007). Undergraduate research experiences support science career decisions and active learning. *CBE-Life Sciences Education, 6*(4), 297-306.

Lopatto, D. (2010). *Science in solution: The impact of undergraduate research on student learning* (1st ed.). Washington, DC: Council on Undergraduate Research.

Lyons, D.L. (2011). *Impact of backwards faded scaffolding approach to inquiry-based astronomy laboratory experiences on undergraduate non-science majors' views of scientific inquiry.* Ph.D. dissertation, University of Wyoming.

Maguire, S. (1998). Gender differences in attitudes to undergraduate fieldwork. *Area, 30*(3), 207-214.

Malina, E.G., and Nakhleh, M.B. (2003). How students use scientific instruments to create understanding: CCD spectrophotometers. *Journal of Chemical Education, 80*(6), 691-698.

Manduca, C., and Kastens, K.A. (2012). Geoscience and geoscientists: Uniquely equipped to study the Earth. *Geological Society of America Special Papers, 486,* 1-12.

Manner, B.M. (1995). Field studies benefit students and teachers. *Journal of Geological Education, 43,* 128-131.

Marques, L., Praia, J., and Kempa, R. (2003). A study of students' perceptions of the organization and effectiveness of fieldwork in earth sciences education. *Research in Science and Technological Education, 21,* 265-278.

Mattox, A.C., Reisner, B.A., and Rickey, D. (2006). What happens when chemical compounds are added to water? An introduction to the Model-Observe-Reflect-Explain (MORE) thinking frame. *Journal of Chemical Education, 83*(4), 622-624.

May, D.B., and Etkina, E. (2002). College physics students' epistemological self-reflection and its relationship to conceptual learning. *American Journal of Physics, 70*(12), 1249-1258.

McConnell, D.A., and van der Hoeven Kraft, K. (2011). Affective domain and student learning in the geosciences. *Journal of Geoscience Education, 59*(3), 106-110.

McConnell, D.A., Jones, M.H., Budd, D.A., Bykerk-Kauffman, A., Gilbert, L.A., Knight, C., Kraft, K.J., Nyman, M., Stempien, J., Vislova, T., Wirth, K.R., Perkins, D., Matheney, R.K., and Nell, R.M. (2009). Baseline data on motivation and learning strategies of students in physical geology courses at multiple institutions: GARNET Part 1, Overview. *Geological Society of America Abstracts with Programs, 41*(7), 603.

McConnell, D.A., Stempien, J.A., Perkins, D., van der Hoeven Kraft, K.J., Vislova, T., Wirth, K.R., Budd, D.A., Bykerk-Kauffman, A., Gilbert, L.A., and Matheney, R.K. (2010). The little engine that could—less prior knowledge but high self-efficacy is equivalent to greater prior knowledge and low self-efficacy. *Geological Society of America Abstracts with Programs, 42*(5), 191.

McCrindle, A.R., and Christensen, C.A. (1995). The impact of learning journals on meta-cognitive and cognitive processes and learning performance. *Learning and Instruction, 5*(2), 167-185.

McKenzie, G.D., Utgard, R.O., and Lisowski, M. (1986). The importance of field trips: A geological example. *Journal of College Science Teaching, 15*, 17-20.

Mestre, J. (2005). *Transfer of learning from a modern multidisciplinary perspective.* Greenwich, CT: Information Age.

Middlecamp, C. (2008). *Chemistry in context: Goals, evidence, gaps.* Paper presented at the National Research Council's Workshop on Linking Evidence to Promising Practices in STEM Undergraduate Education, Washington, DC. Available: http://www7.nationalacademies.org/bose/PP_Middlecamp_WhitePaper.pdf.

Miller, L.S., Nakhleh, M.B., Nash, J.J., and Meyer, J.A. (2004). Students' attitudes toward and conceptual understanding of chemical instrumentation. *Journal of Chemical Education, 81*(12), 1801-1808.

Milliken, K.L., Barufaldi, J.P., McBride, E.F., and Choh, S.J. (2003). Design and assessment of an interactive digital tutorial for undergraduate-level sandstone petrology. *Journal of Geoscience Education, 51*(4), 381-386.

Mogk, D., and Goodwin, C. (2012). Learning in the field: Synthesis of research on thinking and learning in the geosciences. *Geological Society of America Special Papers, 486*, 131-163.

Nagda, B.A., Gregerman, S.R., Jonides, J., Von Hippel, W., and Lerner, J.S. (1998). Undergraduate student-faculty research partnerships affect student retention. *Review of Higher Education, 22*(1), 55-72.

Nakhleh, M.B. (1993), Are our students conceptual thinkers or algorithmic problem solvers? *Journal of Chemical Education, 71*, 52-55.

Nakhleh, M.B., and Mitchell, R.C. (1993). Concept learning versus problem solving: There is a difference. *Journal of Chemical Education, 70*(3), 190-192.

National Research Council. (1999). *How people learn: Bridging research and practice.* M.S. Donovan, J.D. Bransford, and J.W. Pellegrino (Eds.). Committee on Learning Research and Educational Practice, Commission on Behavioral and Social Sciences and Education. Washington, DC: National Academy Press.

National Research Council. (2007). *Taking science to school: Learning and teaching science in grades K-8.* Committee on Science Learning, Kindergarten Through Eighth Grade, R.A. Duschl, H. A. Schweingruber, and A.W. Shouse (Eds.). Board on Science Education, Center for Education. Division of Behavioral and Social Sciences and Education. Washington, DC: The National Academies Press.

National Research Council. (2012). *A framework for K-12 science education practices, cross-cutting concepts, and core ideas.* Committee on a Conceptual Framework for New K-12 Science Education Standards, Board on Science Education. Division of Behavioral and Social Sciences and Education. Washington, DC: The National Academies Press.

Novak, J.D., Gowin, D.B., and Kahle, J.B. (1984). *Learning how to learn.* New York: Cambridge University Press.

Novick, L.R. (1988). Analogical transfer, problem similarity, and expertise. *Journal of Experimental Psychology: Learning, Memory, and Cognition, 14,* 510-520.

Novick, L.R., and Holyoak, K.J. (1991). Mathematical problem solving by analogy. *Journal of Experimental Psychology: Learning, Memory, and Cognition, 17,* 398-415.

Nurrenbern, S.C., and Pickering, M. (1987). Concept-learning versus problem-solving: Is there a difference. *Journal of Chemical Education, 64*(6), 508-510.

Perkins, K.K., Adams, W.K., Pollock, S.J., Finkelstein, N.D., and Wieman, C.E. (2005). Correlating student beliefs with student learning using the Colorado Learning Attitudes about Science Survey. *AIP Conference Proceedings, 790,* 61-64.

Pessoa, L. (2008). On the relationship between emotion and cognition. *Nature Reviews Neuroscience, 9*(2), 148-158.

Petroski, H. (1996). *Invention by design: How engineers get from thought to thing.* Cambridge, MA: Harvard University Press.

Piaget, J. (1978). *The development of thought.* Oxford, UK: Basil Blackwell.

Pickering, M. (1990). Further studies on concept learning versus problem solving. *Journal of Chemical Education, 67*(3), 254-255.

Pintrich, P.R., and De Groot, E.V. (1990). Motivational and self-regulated learning components of classroom academic performance. *Journal of Educational Psychology, 82*(1), 33-40.

Redish, E.F., Steinberg, R.N., and Saul, J.M. (1997). The distribution and change of student expectations in introductory physics. *AIP Conference Proceedings, 399,* 689-698.

Redish, E.F., Steinberg, R.N., and Saul, J.M. (1998). Student expectations in introductory physics. *American Journal of Physics, 66,* 212-224.

Reed, S.K. (1987). A structure-mapping model for word problems. *Journal of Experimental Psychology: Learning, Memory, and Cognition, 13,* 124-139.

Resnick, L., and Klopfer, L. (1989). *Toward the thinking curriculum: Current cognitive research.* Alexandria, VA: Association for Supervision and Curriculum Development.

Rogoff, B., and Wertsch, J.V. (1984). *Children's learning in the zone of proximal development.* San Francisco, CA: Jossey-Bass.

Rosengrant, D., Etkina, E., and Van Heuvelen, A. (2006). An overview of recent research on multiple representations. *AIP Conference Proceedings, 883*(1), 149-152.

Ross, B.H. (1984). Remindings and their effects in learning a cognitive skill. *Cognitive Psychology, 16*(3), 371-416.

Rueckert, L. (2002). Report from CUR 2002: Assessment of research. *Council for Undergraduate Research Quarterly, 22,* 10-11.

Russell, S.H., Hancock, M.P., and McCullough, J. (2007). Benefits of undergraduate research experiences. *Science, 316*(5824), 548-549.

Sadler, T.D., Burgin, S., McKinney, L., and Ponjuan, L. (2010). Learning science through research apprenticeships: A critical review of the literature. *Journal of Research in Science Teaching, 47*(3), 235-256.

Salter, C.L. (2001). No bad landscape. *Geographical Review, 91,* 105-112.

Sandi-Urena, S., Cooper, M.M., and Stevens, R.H. (2011). Enhancement of metacognition use and awareness by means of a collaborative intervention. *International Journal of Science Education, 33*(3), 323-340.

Sandi-Urena, S., Cooper, M.M., and Stevens, R.H. (2012). Effect of cooperative problem-based lab instruction on metacognition and problem-solving skills. *Journal of Chemical Education, 89*(6), 700-706.

Sawrey, B.A. (1990). Concept-learning versus problem-solving—revisited. *Journal of Chemical Education, 67*(3), 253-254.

Schwartz, D.L., Bransford, J.D., and Sears, D. (2005). Efficiency and innovation in transfer. In J. Mestre (Ed.), *Transfer of learning: Research and perspectives* (pp. 1-52). Greenwich, CT: Information Age.

Semken, S. (2005). Sense of place and place-based introductory geoscience teaching for American Indian and Alaska native undergraduates. *Journal of Geoscience Education, 52*(2), 149-157.

Semsar, K., Knight, J.K., Birol, G., and Smith, M.K. (2011). The Colorado Learning Attitudes about Science Survey (CLASS) for use in biology. *CBE-Life Sciences Education, 10*(3), 268-278.

Sere, M.G., Journeaux, R., and Larcher, C. (1993). Learning the statistical analysis of measurement errors. *International Journal of Science Education, 15*(4), 427-438.

Seymour, E., Hunter, A.B., Laursen, S.L., and Deantoni, T. (2004). Establishing the benefits of research experiences for undergraduates in the sciences: First findings from a three-year study. *Science Education, 88*(4), 493-534.

Sibley, D.F. (2009). A cognitive framework for reasoning with scientific models. *Journal of Geoscience Education, 57*(4), 255-263.

Silverthorn, D.U. (2006). Teaching and learning in the interactive classroom. *American Journal of Physiology—Advances in Physiology Education, 30*(4), 135-140.

Simon, H.A. (1978). Information-processing theory of human problem solving. In W.K. Estes (Ed.), *Handbook of learning and cognitive processes* (vol. 5, pp. 271-295). Hillsdale, NJ: Lawrence Erlbaum Associates.

Simpson, R.D., Koballa, T.R., Oliver, J.S., and Crawley, F.E. (1994). Research on the affective dimensions of science learning. In D. Gabel (Ed.), *Handbook of research on science teaching and learning* (pp. 211-234). New York: Macmillan.

Slater, S.J. (2010). *The educational function of an astronomy REU program as described by participating women.* Ph.D. dissertation, University of Arizona, UMI ProQuest 3412598.

Slater, S.J., Slater, T.F., and Shaner, A. (2008). Impact of backwards faded scaffolding in an astronomy course for pre-service elementary teachers based on inquiry. *Journal of Geoscience Education, 56*(5), 408-416.

Snow, R.E. (1989). Toward assessment of cognitive and conative structures in learning. *Educational Researcher, 18*(9), 8-14.

Snow, R.E., Corno, L. and Jackson, D., III. (1996). Individual differences in affective and conative functions. In D.C. Berliner and R. Calfee (Eds.), *Handbook of educational psychology* (pp. 243-310). New York: Macmillan.

Sorby, S.A., and Baartmans, B.J. (2000). The development and assessment of a course for enhancing the 3-D spatial visualization skills of first-year engineering students. *Journal of Engineering Education, 89*, 301-308.

Sousa, D.A. (2011). *How the ELL brain learns.* Thousand Oaks, CA: Corwin Press.

Stains, M., and Talanquer, V. (2008). Classification of chemical reactions: Stages of expertise. *Journal of Research in Science Teaching, 45*(7), 771-793.

Stokes, A., and Boyle, A. (2009). The undergraduate geoscience fieldwork experience: Influencing factors and implications for learning. *Geological Society of America Special Papers, 461*, 291-311.

Storbeck, J., and Clore, G.L. (2007). On the interdependence of cognition and emotion. *Cognition and Emotion, 21*(6), 1212-1237.

Sullivan, W.M. (2005). *Work and integrity: The crisis and promise of professionalism in America.* San Francisco: Jossey-Bass.

Svinicki, M.D. (2004). *Learning and motivation in the postsecondary classroom.* Bolton, MA: Anker.

Svinicki, M.D. (2011). *Synthesis of the research on teaching and learning in engineering since the implementation of ABET Engineering Criteria 2000.* Paper presented at the Second Committee Meeting on the Status, Contributions, and Future Directions of Discipline-Based Education Research. Available: http://www7.nationalacademies.org/bose/DBER_Svinicki_October_Paper.pdf.

Tal, R.T. (2001). Incorporating field trips as science learning environment enrichment—an interpretive study. *Learning Environments Research, 4,* 25-49.

Teichert, M.A., Tien, L.T., Anthony, S., and Rickey, D. (2008). Effects of context on students' molecular-level ideas. *International Journal of Science Education, 30,* 1095-1114.

Tien, L.T., Rickey, D., and Stacy, A.M. (1999). The M.O.R.E. thinking frame: Guiding students' thinking in the laboratory. *Journal of College Science Teaching, 28*(5), 318-324.

Tien, L.T., Teichert, M.A., and Rickey, D. (2007). Effectiveness of a MORE laboratory module in prompting students to revise their molecular-level ideas about solutions. *Journal of Chemical Education, 84*(1), 175-181.

Titus, S., and Horsman, E. (2009). Characterizing and improving spatial visualization skills. *Journal of Geoscience Education, 57*(4), 242-254.

Tobias, S., and Everson, H.T. (1996). *Assessing metacognitive knowledge monitoring.* College Board Report No. 96-1. New York: College Board.

Trigwell, K., Prosser, M., Marton, F., and Runesson, U. (2001). Views of learning, teaching practices, and conceptions of problem solving. In N. Hativa and P. Goodyear (Eds.), *Teacher thinking, beliefs and knowledge in higher education* (pp. 241-254). Dordrecht, The Netherlands: Kluwer Academic.

van der Hoeven Kraft, K.J., Srogi, L., Husman, J., Semken, S., and Fuhrman, M. (2011). Engaging students to learn through the affective domain: A new framework for teaching in the geosciences. *Journal of Geoscience Education, 59*(2), 71-84.

Van Heuvelen, A., and Zou, X. (2001). Multiple representations of work and energy processes, *American Journal of Physics, 69*(2), 184.

Vislova, T., Mcconnell, D., Stempien, J.A., Benson, W.M., Matheney, R.K., van Der Hoeven Kraft, K.J., Budd, D., Gilbert, L.A., Bykerk-Kauffman, A., and Wirth, K.R. (2010). Role of gender in student affect in introductory physical geology courses at multiple institutions. *Geological Society of America Abstracts with Programs, 42*(5).

Vygotsky, L.S. (1978). *Mind in society: The development of higher psychological processes.* Cambridge, MA: Harvard University Press.

Weinstein, C.E., Husman, J., and Dierking, D.R. (2000). Self-regulation interventions with a focus on learning strategies. In B. Monique, R.P. Paul, M. Zeidner, M. Boekaerts, and P.R. Pintrich (Eds.), *Handbook of self-regulation* (pp. 727-747). San Diego, CA: Academic Press.

Yerushalmi, E., Henderson, C., Heller, K., Heller, P., and Kuo, V. (2007). Physics faculty beliefs and values about the teaching and learning of problem solving. I. Mapping the common core. *Physical Review Special Topics—Physics Education Research, 3*(2), 020109-1-020109-31.

Zuckerman, H. (1992). *Scientific elite: Nobel laureates' mutual influences* (2nd ed.). New York: Pergamon Press.

## CHAPTER 8

ABET. (1995). *Vision for change: A summary report of the ABET/NSF/industry workshops.* Baltimore, MD: ABET.

Austin, A.E. (1994). Understanding and assessing faculty cultures and climates. *New Directions for Institutional Research, 84,* 47-63.

Austin, A.E. (1996). Institutional and departmental cultures: The relationship between teaching and research. *New Directions for Institutional Research, 90,* 57-66.

Austin, A.E. (2011). *Promoting evidence-based change in undergraduate science education.* Paper presented at the Fourth Committee Meeting on Status, Contributions, and Future Directions of Discipline-Based Education Research. Available: http://www7.nationalacademies.org/bose/DBER_Austin_March_Paper.pdf.

Austin, A.E., Connolly, M., and Colbeck, C.L. (2008). Strategies for preparing integrated faculty: The Center for the Integration of Research, Teaching, and Learning. In C.L. Colbeck, K.A. O'Meara, and A.E. (Eds.), *Educating integrated professionals: Theory and practice on preparation for the professoriate: New directions for teaching and learning* (pp. 69-81). San Francisco, CA: Jossey-Bass.

Beichner, R.J. (2008). *The SCALE-UP project: A student-centered active learning environment for undergraduate programs.* Paper presented at the National Research Council's Workshop on Linking Evidence to Promising Practices in STEM Undergraduate Education, Washington, DC. Available: http://www7.nationalacademies.org/bose/Beichner_CommissionedPaper.pdf.

Bess, J. (1978). Socialization of graduate students. *Research in Higher Education, 8,* 289-317.

Blackburn. R.T., and Lawrence, J.H. (1995). *Faculty at work: Motivation, expectation, satisfaction.* Baltimore, MD: Johns Hopkins University Press.

Borrego, M., Froyd, J.E., and Hall, T.S. (2010). Diffusion of engineering education innovations: A survey of awareness and adoption rates in U.S. engineering departments. *Journal of Engineering Education, 99*(3), 185-207.

Braxton, J., Lucky, W., and Holland, P. (2002). Institutionalizing a broader view of scholarship through Boyer's four domains. In *ASHE-ERIC Higher Education Report* (vol. 29, no. 2). San Francisco, CA: Jossey-Bass/John Wiley & Sons.

Bronfenbrenner, V. (1979). *The ecology of human development: Experiments by nature and design.* Cambridge, MA: Harvard University Press.

Brower, A., and Inkelas, K. (2010). Living-learning programs: One high-impact educational practice we know a lot about. *Liberal Education 96.*

Bunce, D.M., Havanki, K., and VandenPlas, J.R. (2008). A theory-based evaluation of POGIL workshops: Providing a clearer picture of POGIL adoption. In R.S. Moog and J.N. Spencer (Eds.), *Process-oriented guided inquiry learning (POGIL)* (pp. 100-113). Washington, DC: American Chemical Society.

Clark, S., and Corcoran, M. (1986). Perspectives on the professional socialization of women faculty. *Journal of Higher Education, 57,* 20-43.

Cohen, D.K., and Hill, H. (2001). *Learning policy: When state education reform works.* New Haven, CT: Yale University Press.

Cummings, K. (2008). *The Rensselaer studio model for learning and teaching: What have we learned?* Paper presented at the National Research Council's Workshop on Linking Evidence to Promising Practices in STEM Undergraduate Education, Washington, DC. Available: http://www7.nationalacademies.org/bose/Cummings_CommissionedPaper.pdf.

Cummings, K., Marx, J., Thornton, R., and Kuhl, D. (1999). Evaluating innovation in studio physics. *American Journal of Physics, 67*(7, Suppl. 1), S38-S44.

DeAngelo, L., Hurtado, S., Pryor, J., Kelly, K., Santos, J., and Korn, W. (2009). *The American college teacher: National norms for the 2007-2008 HERI faculty survey.* Los Angeles, CA: Higher Education Research Institute, UCLA.

DeNeef, A.L. (2002). *The Preparing Future Faculty Program: What difference does it make?* Washington, DC: Association of American Colleges and Universities.

Ebert-May, D., Derting, T.L., Hodder, J., Momsen, J.L., Long, T.M., and Jardeleza, S.E. (2011). What is not what we do: Effective evaluation of faculty professional development programs. *Bioscience, 61*(7), 550-558.

Eckel, P., and Kezar, A. (2003). *Taking the reins: Institutional transformation in higher education.* Phoenix, AZ: ACE-ORYX Press.

Fairweather, J. (2005). Beyond the rhetoric: Trends in the relative value of teaching and research in faculty salaries. *Journal of Higher Education, 76,* 401-422.

Fairweather, J. (2008, October). *Linking evidence and promising practices in science, technology, engineering, and mathematics (STEM) undergraduate education: A status report for the National Academies National Research Council Board on Science Education.* Paper presented at the National Research Council's Workshop on Linking Evidence to Promising Practices in STEM Undergraduate Education, Washington, DC. Available: http://www7.nationalacademies.org/bose/Fairweather_CommissionedPaper.pdf.

Fairweather, J., and Paulson, K. (1996). Industrial experience: Its role in faculty commitment to teaching. *Journal of Engineering Education, 85,* 209-216.

Fairweather, J., and Paulson, K. (2005). *The evolution of scientific fields in American universities: Disciplinary differences, institutional isomorphism.* Paper presented at the Annual Conference of the Consortium of Higher Education Researchers, Jyvaskyla, Finland.

Felder, R.M., and Brent, R. (2010). The National Effective Teaching Institute: Assessment of impact and implications for faculty development. *Journal of Engineering Education, 99*(2), 121-134.

Feuer, M.J., Towne, L., and Shavelson, R.J. (2002). Scientific culture and educational research. *Educational Researcher, 31,* 4-14.

Fisher, P.D., Fairweather, J.S., and Amey, M.J. (2003). Systemic reform in undergraduate engineering education: The role of collective responsibility. *International Journal of Engineering Education, 19*(6), 768-776.

Garet, M.S., Porter, A.C., Desimone, L., Birman, B.F. and Yoon K.S. (2001). What makes professional development effective? Results from a national sample of teachers. *American Educational Research Journal, 38,* 915-945.

Gess-Newsome, J., Southerland, S.A., Johnston, A., and Woodbury, S. (2003). Educational reform, personal practical theories, and dissatisfaction: The anatomy of change in college science teaching. *American Educational Research Journal, 40*(3), 731-767.

Gibbs, G., and Coffey, M. (2004). The impact of training of university teachers on their teaching skills, their approach to teaching and the approach to learning of their students. *Active Learning in Higher Education, 5*(1), 87-100.

Handelsman, J., Ebert-May, D., Beichner, R., Bruns, P., Chang, A., DeHaan, R., Gentile, J., Lauffer, S., Stewart, J., Tilghman, S.M., and Wood, W.B. (2004). Education. Scientific teaching. *Science, 304*(5670), 521-522.

Hativa, N. (1995). The department-wide approach to improving faculty instruction in higher education: A qualitative evaluation. *Research in Higher Education, 36*(4), 377-413.

Henderson, C. (2008). Promoting instructional change in new faculty: An evaluation of the Physics and Astronomy New Faculty Workshop. *American Journal of Physics, 76*(2), 179-187.

Henderson, C., and Dancy, M.H. (2007). Barriers to the use of research-based instructional strategies: The influence of both individual and situational characteristics. *Physical Review Special Topics—Physics Education Research, 3*(2), 020102-1–020102-14.

Henderson, C., and Dancy, M.H. (2009). Impact of physics education research on the teaching of introductory quantitative physics in the United States. *Physical Review Special Topics—Physics Education Research*, 5(2), 020107-1–020107-9.

Henderson, C., Beach, A., and Finkelstein, N.D. (2011). Facilitating change in undergraduate STEM instructional practices: An analytic review of the literature. *Journal of Research in Science Teaching*, 48(8), 952-984.

Ho, A., Watkins, D., and Kelly, M. (2001). The conceptual change approach to improving teaching and learning: An evaluation of a Hong Kong staff development programme. *Higher Education*, 42(2), 143-169.

Kezar, A. (Ed.). (2009). *Rethinking leadership in a complex, multicultural, and global environment: New concepts and models for higher education.* Sterling, VA: Stylus.

Krane, K. (2008). *The workshop for new faculty in physics and astronomy.* Presented at the National Research Council's Workshop on Linking Evidence to Promising Practices in STEM Undergraduate Education, Washington, DC. Available: http://www7.nationalacademies.org/bose/Krane_Presentation_Workshop2.pdf.

Lattuca, L., Terenzini, P., Volkwein, F. (2006). *Engineering change: A study of the impact of EC2000.* Baltimore, MD: ABET.

Leslie, D.W. (2002). Resolving the dispute: Teaching is academe's core value. *Journal of Higher Education*, 73(1), 49-73.

Lohmann, J., and Froyd, F. (2010, October). *Chronological and ontological development of engineering education as a field of scientific inquiry.* Paper presented at the Second Meeting of the Committee on the Status, Contributions, and Future Directions of Discipline-Based Education Research, Washington, DC. Available: http://www7.nationalacademies.org/bose/DBER_Lohmann_Froyd_October_Paper.pdf.

Loucks-Horsley, S., Hewson, P.W., Love, N., and Stiles, K.E. (2009). *Designing professional development for teachers of science and mathematics* (3rd ed.). Thousand Oaks, CA: Corwin Press.

Macdonald, R.H., Manduca, C.A., Mogk, D.W., and Tewksbury, B.J. (2004). On the Cutting Edge: Improving learning by enhancing teaching in the geosciences. In *Invention and impact: Building excellence in undergraduate science, technology, engineering, and mathematics (STEM) education.* Washington, DC: American Association for the Advancement of Science.

Macdonald, R.H., Manduca, C.A., Mogk, D.W., and Tewksbury, B.J. (2005). Teaching methods in undergraduate geoscience courses: Results of the 2004 On the Cutting Edge survey of U.S. faculty. *Journal of Geoscience Education*, 53(3), 237-252.

Manduca, C.A., Mogk, D.W., and Stillings, N. (2004). *Bringing research on learning to the geosciences.* Report from a workshop Sponsored by the National Science Foundation and the Johnson Foundation. Northfield, MN: Carleton College, Science Education Resource Center. Available: http://serc.carleton.edu/files/research_on_learning/ROL0304_2004.pdf.

Mazur, E. (1997). *Peer instruction: A user's manual.* Upper Saddle River, NJ: Prentice Hall.

McLaughlin, J., Iverson, E., Kirkendall, R., Bruckner, M., and Manduca, C.A. (2010). *Evaluation report of On the Cutting Edge.* Available: http://serc.carleton.edu/files/NAGTWorkshops/2009_cutting_edge_evaluation_1265409435.pdf.

National Research Council. (2007). *Taking science to school: Learning and teaching science in grades K-8.* Committee on a Conceptual Framework for New K-12 Science Education Standard, Board on Science Education. Division of Behavioral and Social Sciences and Education. Washington, DC: The National Academies Press.

National Research Council. (2009). *A new biology for the 21st century.* Committee on a New Biology for the 21st Century: Ensuring the United States Leads the Coming Biology Revolution. Board on Life Sciences, Division on Earth and Life Studies. Washington, DC: The National Academies Press.

National Science Board. (1986). *Undergraduate science, mathematics and engineering education*. NSB 86010. Arlington, VA: National Science Foundation.

Pfund, C., Miller, S., Brenner, K., Bruns, P., Chang, A., Ebert-May, D., Fagen, A.P., Gentile, J., Gossens, S., Khan, I.M., Labov, J.B., Pribbenow, C.M., Susman, M., Tong, L., Wright, R., Yuan, R.T., Wood, W.B., and Handelsman, J. (2009). Professional development. Summer institute to improve university science teaching. *Science, 324*(5926), 470-471.

Quinn-Patton, M. (2010). *Developmental evaluation: Applying complexity concepts to enhance innovation and use*. New York: Guilford Press.

Rogers, E.M. (2003). *Diffusion of innovations* (5th ed.). New York: The Free Press.

Sawada, D., Piburn, M.D., Judson, E., Turley, J., Falconer, K., Benford, R., and Bloom, I. (2002). Measuring reform practices in science and mathematics classrooms: The reformed teaching observation protocol. *School Science and Mathematics, 102*, 245-253.

Schuster, J.H., and Finkelstein, M.J. (2006). *The American faculty: The restructuring of academic work and careers*. Baltimore, MD: Johns Hopkins University Press.

Seymour, E. (2001). Tracking the process of change in U.S. undergraduate education in science, mathematics, engineering, and technology. *Science Education, 86*, 79-105.

Silverthorn, D.U. (2006). Teaching and learning in the interactive classroom. *American Journal of Physiology—Advances in Physiology Education, 30*(4), 135-140.

Smith, B., MacGregor, J., Matthews, R., and Gabelnick, F. (2004). *Learning communities: Reforming undergraduate education*. San Francisco, CA: Jossey-Bass.

Sorensen, C.M., Churukian, A.D., Maleki, S., and Zollman, D.A. (2006). The new studio format for instruction of introductory physics. *American Journal of Physics, 74*(12), 1077-1082.

Tobias, S. (1992). *Revitalizing undergraduate science: Why some things work and most don't*. An occasional paper on neglected problems in science education. Tucson, AZ: Research Corporation.

U.S. Department of Education. (2005). *2004 National Survey of Postsecondary Faculty (NSOPF:04) report on faculty and instructional staff in fall 2003*. Washington, DC: National Center for Education Statistics.

Wieman, C., Perkins, K., and Gilbert, S. (2010). Transforming science education at large research universities: A case study in progress. *Change: The Magazine of Higher Learning, 42*(2), 7-14.

Wilson, S.M. (2011, May). *Effective STEM teacher preparation, instruction, and professional development*. Paper presented at the workshop of the National Research Council's Committee on Highly Successful Schools or Programs for K-12 STEM Education, Washington, DC. Available: http://www7.nationalacademies.org/bose/STEM_Schools_Wilson_Paper_May2011.pdf.

Wlodkowski, R.J. (1999). *Enhancing adult motivation to learn: A guide to improving instruction and increasing learner achievement* (revised edition). San Francisco, CA: Jossey-Bass.

Wood, W.B., and Gentile, J.M. (2003a). Teaching in a research context. *Science, 302*(5650), 1510.

Wood, W.B., and Gentile, J.M. (2003b). Meeting report: The first National Academies Summer Institute for Undergraduate Education in Biology. [Congresses]. *Cell Biology Education, 2*(4), 207-209.

Yerushalmi, E., Cohen, E., Heller, K., Heller, P., and Henderson, C. (2010). Instructors' reasons for choosing problem features in a calculus-based introductory physics course. *Physical Review Special Topics—Physics Education Research, 6*(2), 020108-1–020108-11.

# CHAPTER 9

Fensham, P.J. (2004). *Defining an identity: The evolution of science education as a field of research*. Boston, MA: Springer.

Huffman, D., and Heller, P. (1995). What does the Force Concept Inventory actually measure? *Physics Teacher, 33*(2), 138-143.

National Research Council. (2006). *Learning to think spatially: GIS as a support system in the K-12 curriculum*. Committee on Support for Thinking Spatially: The Incorporation of Geographic Information Science Across the K-12 Curriculum. Geographical Sciences Committee. Board on Earth Sciences and Resources, Division on Earth and Life Studies. Washington, DC: The National Academies Press.

# Appendix

# Biographical Sketches of Committee Members and Staff

**SUSAN R. SINGER** (*Chair*) is the Laurence McKinley Gould professor of natural sciences in the Biology Department at Carleton College. Her research focuses on the development and evolution of flowering in legumes and on undergraduate learning of genomics. She has directed Carleton's Perlman Center for Learning and Teaching and chaired the Biology Department. Previously, she was a National Science Foundation (NSF) program officer in developmental mechanisms. A recipient of the Excellence in Teaching Award from the American Society of Plant Biology (ASPB), she is also a fellow of the American Association for the Advancement of Science. She coauthors an introductory biology text and is editor of the ASPB's plant biology education book series. She is currently serving on the advisory board of the Education and Human Resources Division of the NSF, the board of the ASPB Education Foundation, and the board of the iPlant Cyberinfrastructure Collaborative board. She has served on several National Research Council (NRC) study committees and was a member of the NRC's Board on Science Education. She holds a Ph.D. in biology from Rensselaer Polytechnic Institute.

**ROBERT BEICHNER** is a member of the Physics Education Research and Development Group at North Carolina State University. He is also the director of the university's STEM Education Initiative, with a mission to study and improve STEM education from "K to gray" in North Carolina and around the world. His research addresses student learning and improving physics education. His largest current project is the creation and study of a learning environment supporting a new way to teach: Student-Centered

257

Active Learning Environment for Undergraduate Programs—SCALE-UP—an approach that has been adopted at more than 50 institutions. For his education reform efforts, he was named the 2009 North Carolina professor of the year by the Council for Advancement and Support of Education and the 2010 national undergraduate science teacher of the year by the Society of College Science Teachers. He is the founding editor of a journal of the American Physical Society, *Physical Review Special Topics: Physics Education Research*. He holds a Ph.D. in science education from The State University of New York, Buffalo.

**STACEY LOWERY BRETZ** is a professor of chemistry at Miami University (Ohio). Previously, she was on the faculty of the University of Michigan–Dearborn and of Youngstown State University. Her current research relates to the assessment of student learning, including chemistry concept inventories, the application of cognitive science theories and qualitative methodologies to chemistry education research, inquiry in the laboratory, and children and chemistry. With support from the National Science Foundation, she has created a series of conferences for chemistry education research graduate students. She currently serves as chair of the board of trustees for the American Chemical Society Division of the Chemical Education Examinations Institute, and she is a fellow of the American Association for the Advancement of Science. She is a recipient of the E. Phillips Knox award for undergraduate education from Miami University and of both the distinguished professor of teaching award and the research awards from Youngstown State University. She holds a B.A. from Cornell University and an M.S. from Pennsylvania State University, both in chemistry, and a Ph.D. in chemistry education research from Cornell.

**MELANIE COOPER** is an alumni distinguished professor of chemistry at Clemson University. Her research has investigated problem solving in a wide variety of areas, including laboratories and large enrollment lectures. Her work on methods to assess and improve students' problem-solving abilities and strategies has focused on interventions that promote metacognitive activity. An outgrowth of this research is the development and assessment of evidence-driven, research-based curricula. She is a fellow of the American Association for the Advancement of Science and has received a number of awards for excellence in teaching. She holds a B.S., an M.S., and a Ph.D. from the University of Manchester (England).

**SEAN DECATUR** is dean of the College of Arts and Sciences and a professor of chemistry at Oberlin College. Previously, he served as associate dean of faculty for science and the Marilyn Dawson Sarles professor of life sciences and professor of chemistry at Mount Holyoke College. His primary field of research is in the area of protein structure and protein folding. His interests also include the field of science studies, in particular the intersection of race and science in the United States. He has taught a wide range of courses in chemistry, including introductory chemistry, physical chemistry, and biophysical chemistry, and he has mentored more than 50 undergraduate students on research projects. He has received several national awards, including a National Science Foundation Faculty Early Career Development (CAREER) Program grant. He received a B.A. from Swarthmore College and a Ph.D. in biophysical chemistry from Stanford University.

**JAMES FAIRWEATHER** is the Mildred B. Erickson distinguished chair in higher, adult, and lifelong education at Michigan State University, where he also directs the Center for Higher and Adult Education. His works focuses on faculty roles and rewards, reform in undergraduate STEM education, the globalization of higher education policy, and the role of higher education in economic development. His research has been funded by private and nonprofit organizations, as well as the Dutch and Omani governments and the National Science Foundation (NSF). Most recently, he has been co-principal investigator of the NSF-funded Center for the Integration of Research, Teaching and Learning. He has chaired the editorial board of the *Journal of Higher Education*. He received the exemplary research career award from a division of the American Educational Research Assocation, and he has been a Fulbright scholar and held an Erasmus Mundus professorship from the European Union. He received a Ph.D. in higher education from Stanford University.

**MARGARET L. HILTON** (*Senior Program Officer*) has directed and contributed to a wide range of studies at the National Research Council, including those on high school science laboratories, the role of state standards in K-12 education, foreign language and international studies in higher education, international labor standards, and the information technology workforce. Prior to joining the National Research Council staff, she was a consultant to the National Skill Standards Board and she directed studies of workforce training, work reorganization, and international competitiveness at the Office of Technology Assessment. She holds a B.A. in geography from the University of Michigan, an M.A. in regional planning from the University of North Carolina at Chapel Hill, and an M.A. in human resource development from George Washington University.

**KENNETH HELLER** is a professor of physics at the University of Minnesota. His research in high-energy particle physics focuses on the properties of neutrino oscillations. He has conducted studies of quark dynamics from strong interactions of hadrons, quark confinement from magnetic moments of baryons and their weak decay properties, and muons from high energy interactions. He is also actively involved in research in physics education, and he has served as the president of the American Association of Physics Teachers. He leads a physics education research group that is investigating better ways to teach problem solving through the use of cooperative groups, context-rich problems, and expert strategies. As part of this work, he is developing techniques to assess problem solving in physics. He received his B.A. from the University of California and a Ph.D. in physics from the University of Washington in Seattle.

**KIM KASTENS** is a Doherty senior research scientist at the Lamont-Doherty Earth Observatory and director of Columbia University's program in earth and environmental science journalism, both at Columbia University. Her early work in marine geology focused on focused on mapping the seafloor and interpreting the tectonic and sedimentary processes that shaped it. More recently, she shifted her focus to geoscience education, learning science research, and instructional technology, particularly at the Ph.D. level. Her research interests include exploration of children's map skills, use of maps to communicate with policy makers, and visualization of three-dimensional structures by scientists and geoscience students. She designed and produced *Where Are We?*, an educational software package and associated curricula for elementary school children. She served on the National Research Council's Committee on the Review of the National Oceanic and Atmospheric Administration's Education Program. She holds a B.A. in geology and geophysics from Yale University and a Ph.D. in oceanography from the Scripps Institution of Oceanography at the University of California, San Diego.

**MICHAEL E. MARTINEZ** was a professor in the Department of Education at the University of California (UC) at Irvine, where he also served as codirector of the university's joint doctoral program in education with California State University and as vice chair of the Department of Education. Before joining the UC Irvine faculty, he worked at the Division of Research at the Educational Testing Service in Princeton, New Jersey, where he developed new forms of computer-based testing for assessment in science, architecture, and engineering and as a program director at the National Science Foundation. Earlier in his career, he was a high school science teacher. His research interests were learning and cognition, intelligence, and science and mathematics education. He served on several National Research Council study committees. His honors include the presidential commendation for contributions to psychology from the American Psychological Association. He received a Ph.D. in educational psychology from Stanford University.

**DAVID MOGK** is a professor of geology at Montana State University, and he is the co-principal investigator of the university's image and chemical analysis laboratory. His research interests in geology include the evolution of ancient (> 2.5 billion years old) continental crust in southwest Montana, petrologic processes in the mid-crust, spectroscopy of mineral surfaces, and the search for life in extreme environments (from Yellowstone hot springs to the Lake Vostok ice core). He is actively involved in education research and innovation. With support from the National Science Foundation (NSF), he recently worked on the development of the digital library for earth system education and the National Science Digital Library, and he is currently working on projects related to geoscience education. He is currently a member of NSF's EarthScope Science and Education Advisory Board. He is a recipient of the excellence in geophysical education award of the American Geophysical Union. He received a B.S. in geology from the University of Michigan and an M.S. and a Ph.D. in geology from the University of Washington.

**NATALIE R. NIELSEN** (*Study Director*) is a senior program officer with the National Research Council's Board on Science Education, where she has also worked on other studies related to K-12 STEM education. Before joining the National Research Council, she was the director of research at the Business-Higher Education Forum, where her work focused on college readiness, access, and success, particularly in STEM, and a senior researcher at SRI International, where she conducted evaluations of a wide variety of reform efforts, including technology initiatives, after-school programs, teacher quality, data-driven decision making, youth development programs, and high school reform. She has also served as a staff writer for Project 2061 of the American Association for the Advancement of Science, exhibit researcher at the National Museum of Natural History, and exhibit writer and internal evaluator at the San Diego Natural History Museum. She holds a B.S. in geology from the University of California at Davis, an M.S. in geological sciences from San Diego State University, and a Ph.D. in education from George Mason University.

**LAURA R. NOVICK** is an associate professor in the Department of Psychology and Human Development in the Peabody College of Education and Human Development at Vanderbilt University. Her current research explores issues at the interface of cognitive psychology and evolution education and has influenced how tree-of-life diagrams are depicted in biology textbooks. In connection with this work, she recently participated in an interdisciplinary project with natural history museums to make recommendations for improving their tree-of-life exhibits. She has previously conducted research in areas such as analogical problem solving, expertise, and diagrammatic reasoning. She currently serves on the advisory board for an engineering and robotics education project at Georgia Institute of Technology. She is a fellow of the Association for Psychological Science and a recipient of a Spencer fellowship from the National Academy of Education. She holds a B.S in psychology from the University of Iowa and a Ph.D. in cognitive psychology from Stanford University.

**MARCY OSGOOD** is associate professor and vice chair of education in the Department of Biochemistry and Molecular Biology at the University of New Mexico. She also serves as a curriculum developer and faculty member for the university's premedical enrichment program, a post-baccalaureate program for educationally disadvantaged students preparing to enter medical school. One aspect of her work has been putting into practice numerous multicontextual learning and teaching modalities for minority students. She is a mentor to other university faculty in curriculum development/course design in conjunction with the university's School of Medicine teacher and educational development. Previously, she was at the University of Michigan in Ann Arbor, where she coordinated a personalized system of instruction program in biochemistry and taught biology to both majors and nonmajors in the field. She has served as the director and outreach director, respectively, for two New Mexico programs, the Southwest Graduate Coalition Bridges to the Doctorate and the New Mexico Idea Network of Biomedical Research Excellence. She received a Ph.D. in biology from Rensselaer Polytechnic Institute.

**HEIDI A. SCHWEINGRUBER** (*Report Co-Editor*) is the deputy director of the Board on Science Education at the National Research Council (NRC), where she has directed or co-directed several studies on K-12 science education, including the project that resulted in the NRC report, *A Framework for K-12 Science Education* (2012). Prior to joining the NRC, she was a senior research associate at the Institute of Education Sciences in the U.S. Department of Education; was the director of research for the School Mathematics Project at Rice University, an outreach program in K-12 mathematics education; and taught in the psychology and education departments at Rice University. She has a Ph.D. in psychology (developmental) and anthropology, and a certificate in culture and cognition from the University of Michigan.

**TIMOTHY F. SLATER** is a professor at the University of Wyoming where he holds the Wyoming excellence in higher education endowed chair of science education. As part of the university's Center for Astronomy & Physics Education Research (CAPER), his scholarship focuses on the cognitive processes underlying how students engage in learning science and how teachers learn to teach science. He works with college and university faculty members on improving teaching practices for both nonscience majors and future teachers. He has been an elected board member of the National Science Teachers Association, the Society of College Science Teachers, the American Astronomical Society, and the Astronomical Society of the Pacific. He was also a founding member of the editorial board for the Astronomy Education Review. He holds a B.S. in physical science and a B.S. in secondary education from Kansas State University, an M.S. in physics and astronomy from Clemson University, and a Ph.D. in geological sciences and geophysics from the University of South Carolina.

**KARL A. SMITH** is cooperative learning professor of engineering education at the School of Engineering Education at Purdue University. He also has appointments as the Morse-alumni distinguished teaching professor and a professor of civil engineering at the University of Minnesota. His research and development interests include building rigorous research capabilities in engineering education; the role of cooperation in learning and design; problem formulation, modeling, and knowledge engineering; and project and knowledge management. He is a fellow of the American Society for Engineering Education and past chair of the society's Educational Research and Methods Division. He has served as the principal investigator or co-principal investigator on several National Science Foundation projects, including two centers for learning and teaching and a dissemination project on course, curriculum, and laboratory improvement. He holds a B.S. and an M.S. in metallurgical engineering from Michigan Technological University and a Ph.D. in educational psychology from the University of Minnesota.

**WILLIAM B. WOOD** is a distinguished professor of molecular, cellular, and developmental biology (emeritus) at the University of Colorado, Boulder. Previously, he was on the faculty of the California Institute of Technology. His early research focused on the assembly of complex viruses that infect bacteria. More recently, his research interests have included biology education and the genetic control and molecular biology of axis formation, pattern formation, and sex determination in development of the nematode *Caenorhabditis elegans*. He is the recipient of several awards for his scientific achievements, including the Bruce Alberts award for distinguished contributions to science education from the American Society for Cell Biology. He is a member of the National Academy of Sciences and has served on National Research Council study committees. He received a Ph.D. in biochemistry from Stanford University.